热带度假村设计
HOTEL AND RESORT DESIGN

热带度假村设计

HOTEL AND RESORT DESIGN

[泰] 哈比塔事务所 编著
齐梦涵 译

广西师范大学出版社
·桂林·
images
Publishing

图书在版编目(CIP)数据

热带度假村／泰国哈比塔事务所编著；齐梦涵译．—桂林：广西师范大学出版社，2018.5
ISBN 978-7-5598-0515-7

Ⅰ．①热… Ⅱ．①泰… ②齐… Ⅲ．①饭店-建筑设计-世界-图集 Ⅳ．①TU247.4-64

中国版本图书馆 CIP 数据核字(2018)第 055061 号

出 品 人：刘广汉
责任编辑：肖　莉
助理编辑：齐梦涵
版式设计：吴　迪

广西师范大学出版社出版发行
(广西桂林市五里店路 9 号　　邮政编码：541004
网址：http://www.bbtpress.com)
出版人：张艺兵
全国新华书店经销
销售热线：021-65200318　021-31260822-898
恒美印务(广州)有限公司印刷
(广州市南沙区环市大道南路 334 号　邮政编码：511458)
开本：635mm×965mm　1/8
印张：32　　字数：30 千字
2018 年 5 月第 1 版　　2018 年 5 月第 1 次印刷
定价：258.00 元

目录
CONTENTS

序

FOREWORD

哈比塔事务所已经在此行业从业35年。正如本书内容所展示的那样，我们的工作重心一直在度假村项目上，且作品最初只见于泰国，不过现在已遍布整个亚洲、中东与中欧。

我们总是从基本原则的角度出发，审视每个项目，再开发出可以使顾客体验当地自然或人文环境的概念。度假村的地点决定它是会坐落在雨林谷的某一侧，还是停在一个环礁湖上，抑或是退居于一片宁静的稻田里。

我们在酒店行业的工作总会不断面临各种挑战，这其中也不乏文化方面的。我们所雇佣的员工主要来自于两种建筑流派——他们其中一部分学习泰国传统建筑，其他人学的则是现代主义建筑。这种人员构成对于泰国的建筑实践其实是不太常见的，而在我们的办公室中，这两种流派的建筑师之间的协同工作使我们能够做到既考虑历史传承，又有意识地考虑创新方法。

当然，如果没有对我们的工作十分支持的客户和颇具才干的承包商，还有他们之间融洽的合作关系，本书中所展现的我们在过去40年中所做的设计成果就不会化为现实。我们希望把本书献给他们，以及这些年来与我们有过合作的各位顾问和设计师。

科瑞斯塔·罗查纳科恩

前言

INTRODUCTION

唐考·潘宁，副教授

在21世纪的第二个十年中前行，建筑设计遇到了许多挑战，这其中的紧迫问题不仅有环境问题，还有不少以建筑实践在文化、社会经济或美学等方面为基本前提的问题，使建筑可以从多种不同的角度被加以分析。虽然我们可以争辩建筑具有自主性，其形式与几何可以被独立地理解，但是我们也可以否认这种所谓的自主性，去探索范围更广的文化条件所产生的影响，无论这种影响是来自技术、社会还是经济。在从传统到现代的范围之内还存在各种不同的思维方式和实践架构，而这些理论则将在未来几十年中改变我们的生活环境。

在传统建筑根深蒂固的地方，往昔的意象不可避免地变成了一种修辞，人们对它没有清晰的认识。然而，当我们思考目前建筑的情况时，我们会发现与此相反的看法也是正确的。在许多地方，人们能明确理解传统建筑的定义，而现代建筑却会让他们摸不着头脑。在过去的50年里，泰国现代建筑一直热衷于寻找自己的定义与身份。是坚持传统的形式与外表吗？是采用新材料与新技术吗？是应对气候与环境变化吗？是接受外国的影响吗？如果把这些问题都考虑进去，那么现在的问题就变成了我们要如何去做。一代又一代的建筑师对此进行了各种尝试，进而提出了迥然不同的解决方案。但是，对泰国现代建筑的身份认同依然如同水中之月。

在过去的40年中，许多泰国建筑设计公司都采取了各种方式去阐释热带及现代泰国建筑，哈比塔事务所也是其中之一。多年来，哈比塔事务所已经完成了许许多多规模各异的设计项目。尽管这些项目各不相同，却都被定义于度假旅店的范畴之中。除了满足每个项目的具体要求，建筑师们也在不断探寻一种不仅适用于泰国，也适用于其他地区的当代解决方案。哈比塔事务所的建筑作品与其说是外观独具特色，不如说是建筑物之间的位置关系是其特点。建筑从来就不是仅为取悦眼睛而建造的，它与我们的生活还有人为创造及自然环境息息相关。他们的建筑就好像是要表达某些关于空间和时间的最基本的概念一样。这不体现在建筑风格上，因为所有项目的建筑风格都各不相同，它体现在建筑与对空间和结构的布局有关的文脉结合上，正是这种结合使得他们的建筑作品与众不同。

对于哈比塔事务所来说，为重新定义作为更广阔视野的组成部分的建筑构造而谈论建筑所处的语境，意味着我们先要重新思考建筑物与其空间和物质环境的联系，这里的物质环境既包括自然地形也包括文化情景。在这个意义上，建筑的时代性不意味着某个时代的建筑必然要与此前的时代背景完全分离，建筑结构本身也不应被视为一个完全独立的整体，而应被纳入包括其在内的物质和非物质语境，作为其下的一个活跃的组成部分。

对于20世纪早期的热带建筑来说，人地关系曾是开放的，人们与其居住的地方虽然置于一个开放的平台之上，却依然受到保护。在每个实例中，人们都可以看出对于建筑来说什么是基本的，主要的，可是到了20世纪末，我们对自然环境的驯化变成简单把我们自己围在玻璃围墙之中。在过去的 20 年里，人们意识到气候与环境变化所产生的问题，无论是在泰国还是其他地方，人们都逐渐开始对乡土和传统建筑产生兴趣。哈比塔事务所的这种工作取向不仅仅是出于对回归重建传统及乡土建筑的愿望，虽然他们的做法与那些由此主旨及意图产生的活动十分相近。通过专注于平淡无奇的情况与构造，他们的目标也包括将幻想表现出来，并与日常实践相结合。这种平衡不在于对理想中的传统及乡土文化的崇敬，而在于对变革的接受。为了使建筑结构能够立于其自身所处的既非开始也非结束的地域与时代之中，我们需要对过去与现在进行建构性解释。居住于此的人们或许能将祖先追溯到 200 年前，但是对他们来说更加重要的是活在当下，即生活在现在的气候需求、功能需求与文化需求之中，那么必将有建筑为了满足这些人的需求而存在。

在这样的思维框架内，哈比塔事务所的设计过程往往关注建筑的实际建设，而不会开始并结束于绘图板。设计必须从对生活和环境的调查开始。灵感从不会脱离大背景而存在，它常常来自于某个地方平淡——有时甚至是渺小——的方面。除了每个项目的纲领性框架外，建筑师们也要考虑每个项目当地固有的材料、劳动力、施工方法和工艺。建筑物的魅力不仅体现在空间和形式上，也体现在材料和工艺方面。当材料的功能超出其自身时，建筑也会为人们提供超越其自身的体验。

这种设计倾向使哈比塔事务所能够为不同地点的项目营造出不同类型却互有关联的环境与语境关系。每个项目的设计都经过了对施工地当地习惯的仔细考虑和理解。除了纲领性和风格上的差异，所有的项目均采用有意识地与周围环境进行互动的设计方法。设计师对建筑与景观的整合不仅意味着两者在物质层面具有相似之处，也意味着两者之间存在对比，即使在外观上不相互匹配，却在内含有时空延续，这种延续为建筑与其所处传统及乡土环境之间的关系提供一个开放式框架。通过建筑物整合景观不仅限于外观，还包括时间和空间中的光线、气候、神韵，建筑进入并改造这个大环境，成为其中的一部分。它与补全周围景观的建筑形成鲜明对照，但也展示出联系的不同表达——永远不要将建筑物自身定义为整体，因为建筑从来不会脱离其地点而存在。

除了重新定义周围的景观，哈比塔事务所的建筑也试图回应每个项目当地的局部气候。对于热带地区来说，空间及其周边的围护结构不仅共同创造出人员流通的路径，也创造了光和空气在流动方面的关联性。从这个意义上来说，建筑的外壳和表皮远不止体现风格和具有装饰性，它们也在功能、空间和气候层面产生作用。尽管每个项目在风格上都有差异，但空间总是被墙壁包围着的，这些墙壁不受传统意义上的建筑墙面的定义所限制，相反，其更具多样性。除了标定空间，围护结构让建筑物呼吸、投下阴影，或使光线照射进来。在哈比塔事务所的设计中，空间的联系与相互关联的次序是独特的，其效果则是人们可以在不知不觉中从被围住的空间移动到开放的空间之中。建筑及其周围的环境相互融合，不单表示他们在物理层面有相似之处，还表示它们已经自然地成为彼此的一部分。

建筑的特性从不局限于它的墙壁，它周围的景观往往可以使其自身更加丰富。土地和建筑物的关系不是物质的简单相加，人们更希望在其中看见和谐。

哈比塔事务所的作品常保留人工与自然相互碰撞的痕迹，而非试图消除建筑和环境之间的差异，因为人们可以通过设计来清晰表述建筑物与接纳它的景观之间的具有建设性的

对话。这表明被精确定义的建筑要素可以被集成到自然环境中去，并依然保持自己的特性。为了理解这些设计，我们必须超越仅把它们当作视觉上赏心悦目的对象这种观念，进一步关注其在公共及私人这两方面的功能。我们可以从建筑物与其环境的衔接之中，重新定义和解释隐私与围护结构这两者概念上的关联。只要每个场景都以合适的朝向，按照恰当的顺序排列，并被安置在合适的地点，那么私人空间也可以与公共空间一样开放，与景观相连。建筑、景观和气候不是相互独立的主题，而是一个由复杂的语汇和概念紧密相连的复合体，是展示它们如何在不同的场所被连接在一起的。

从这个意义上说，设计取决于对情景的相似性与差异性的洞察，它必须为情景之间的相互关系奠定基础，而不能只聚焦于设计对象的形态或形式。传统意义上的建筑身份被替代，结果产生出一种具有流动性但依然界定清晰的建筑架构。因此，这些建筑从设计师的个人风格角度来说并不具有典型性，但它们在实现现实与理念的统一，在独特环境中进行布局方面却十分有意义。这些设计也对环境的变化有很好的兼容性，它们能很好地满足功能框架上的需求，也为将来可能出现的改造留出了足够的空间。

这些作品表现出了自然环境对建筑的强大影响力，但是它们也展现出我们的建筑对环境的重塑能力。这些建筑也许不是我们对当代设计进行探索的终点，但是它们依然体现了建筑是怎样对今天这个时代做出回应的。哈比塔事务所的建筑实践特征是：对当代区域有认同感、受气候制约、关注地形的特殊性，因此建筑对当地的影响程度也成为他们关注的一个中心问题。哈比塔事务所的建筑师们的任务不是设计出新颖的建筑形态，而是探索负责任的创新建筑方法。地点本身为我们提供了无限的可能，并使有意义的设计决策能够产生，进而允许建筑能够与它们整合在一起，共同成长。建筑既不沉默也不喧闹，它们能在其所处的地方找到自己的语言，在那里崛起，梳理其存在的语境。建筑物不再是理念的中心，它与自身外延的理念的重新组合有关——它们共同成为地域文脉的一部分。

我们一旦把这些不同的因素都考虑进来，描述一个地点的任务就变得更加清晰了。发展语言和概念能够证明相隔遥远、差别明显的因素之间是如何相互联系的，而且当地传统和新的愿景、局部气候和新技术、地形条件和建筑干预、地域文化和全球需求，这些相互分离的事情也是可以联系在一起的。从这层意义上来说，居住指的就是关注建筑所具有的不同性能，并发现它们所具有的相似之处。建筑不应是令人觉得遥远的，而应带给人熟悉感，能够体现出不同地区的典型情况。当我们不再用传统与创新、功能与审美、地方条件和全球要求这种二分法来看待建筑，我们会发现建筑与环境和社会的联系是不断发展和变化的。正是这些联系，建筑才能够克服其自身的平凡，将自身转化为理想的状态。

酒店与度假村

HOTELS AND RESORTS

宁凡湾六善酒店

SIX SENSES NINH VAN BAY

Location Nha Trang, Vietnam
Date 2004
Interior designer Six Senses Creative Department

Situated in natural isolation across the bay from the resort city of Nha Trang, this project permitted research into two areas of intense interest to this office. Firstly, developing an architectural response to a site's essential nature. And secondly, developing a fitting way to develop the site, particularly when local craftsmen are involved.

Impressive rock formations and boulders dominate the site. Instead of the original brief to develop water villas here, we proposed rock villas as the key selling point of the resort. The interface between building and nature is seen in a variety of ways across the site—the integration of the spa with waterfall, and the integration of reception, restaurant, and rock villas directly with the rocks themselves.

As the development of the project progressed, a method of design and construction emerged that was quite specific to this project. This was characterized by close engagement with Vietnamese craftsmen and the use of materials readily available on the site (such as rock and roof thatching). In addition, the requirement for partial prefabrication of some buildings added another layer to the unique way this project was delivered.

Six Senses Ninh Van Bay consists of 51 units, comprising 31 beach villas, five water villas, five rock villas, five hilltop villas and five spa suite villas.

The reception building's design is inspired by the traditional Vietnamese 'Dinh' or village communal house, an ideal precedent for the informal reception environment sought.

The architecture of the Dinh is reinterpreted in public buildings across the resort—its timber structural details examined and simplified in both building structure and bespoke furniture. Traditional Vietnamese carpentry eschews the use of nails, instead fixing tenon and mortise joints with timber pins.

As well as rock and timber, the palette of natural materials included very distinctive thatch roofing. From the underside, the very beautiful pattern of stems fixed to battens is exposed for the guests to see.

We were greatly privileged on this project to travel around Vietnam to better understand vernacular buildings. We are proud to help disseminate the intellect of Vietnamese timber craftsmanship, in a modern architectural development with modern functional demands.

MASTER PLAN
1 Arrival jetty
2 Lobby
3 Dining by the Pool
4 Gallery courtyard
5 Wine Cave
6 Sub Club—library
7 Drinks by the Bay
8 Dining on the Jetty
9 Dining by the Bay
10 Spa
11 Beachfront Pool Villa
12 Rock Pool Villa
13 Pool Villa with Rockery
14 Hill Top Pool Villa
15 Back-of-house & host accommodations
0 10 50 100 200m
N

DRINKS & DINING BY THE BAY
1 Drinks by the Bay
2 Game room
3 Dining by the Bay
4 Kitchen
0 0.5 2.5 5 10m
1
2
3
4

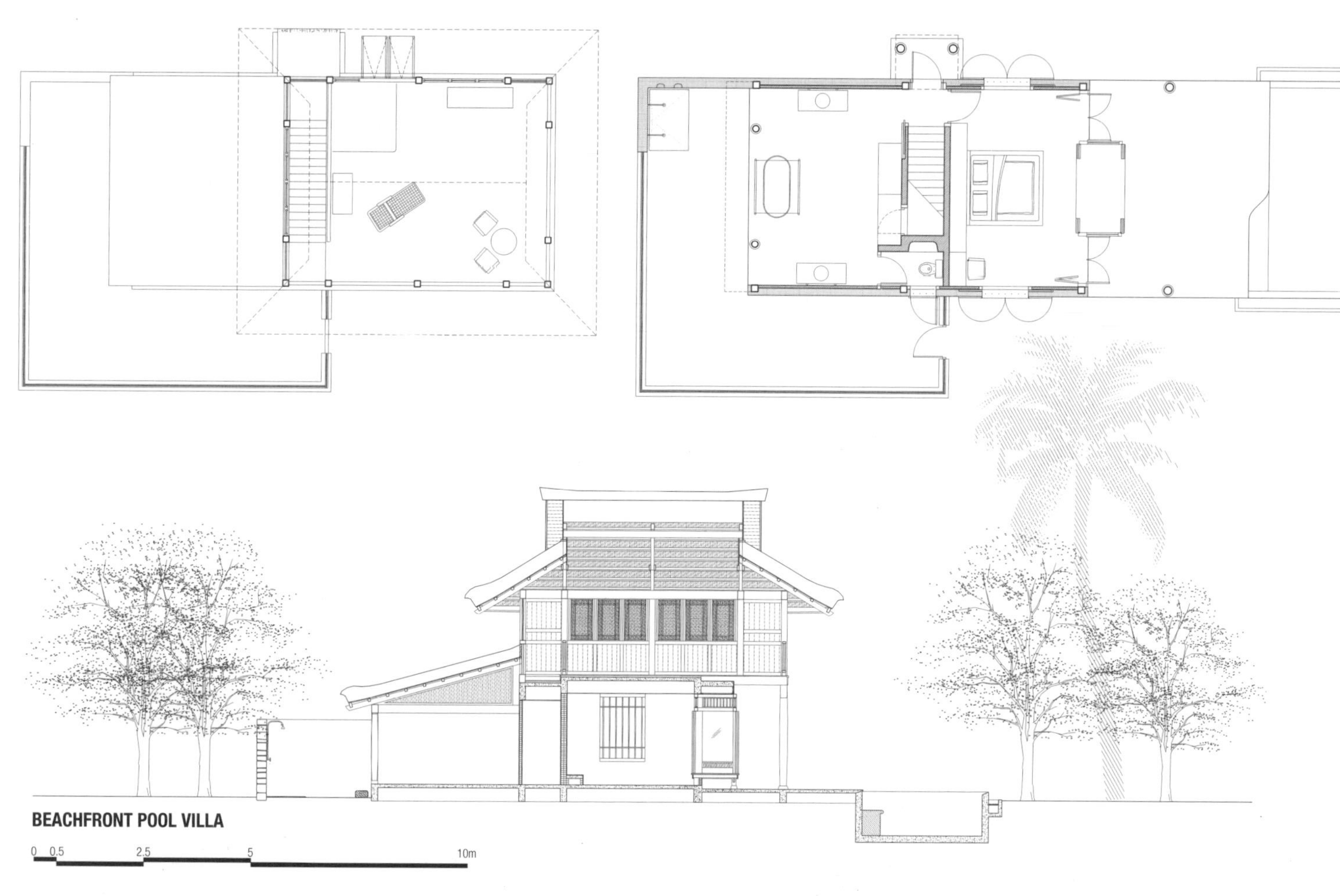
BEACHFRONT POOL VILLA
0 0.5 2.5 5 10m

ROCK POOL VILLA

0 0.5 2.5 5 10m

华欣爱梵森酒店

SIX SENSES HUA HIN

The Six Senses Hua Hin, currently the Sheraton Hua Hin Pranburi Villas, are a development of 53 pool villas located on a small, secluded estate on beach near Huahin, Thailand. The land on which the resort is situated was previously a coconut plantation. The concept of the project was to keep as many of the existing palms as possible and create the atmosphere of a village in the coconut farm.

A curving path leads from the road entrance down the center of the site to the beach. On each side of the path are clustered the villas, located behind high privacy walls with a free-flowing form. An important feature is the plaster used as the surface finish of these walls. Its rough texture results from using cut straw mixed with the cement, which was then applied liberally by hand. The uneven finish aims to naturally soften the appearance of the walls, and creates a unique feel to this project, from both the public areas and from each villa's private garden. The high walls afford privacy for each villa's landscaped pool area, as well as bedroom and bathroom.

Located in the center of this path to the sea, near the entrance to the resort, is the main reception building and restaurant. These overlook a large lotus pond, in which are located three sala. These are sunken into the pond, so that the water level is equal to the top of the sofas on which guests can dine. The third sala provides a stage for entertainment—a focal point for the restaurant and pond area. Indeed, this space is really the heart of the village.

Location Prachuap Khiri Khan, Thailand
Date 2004
Interior designer Six Senses Creative Department

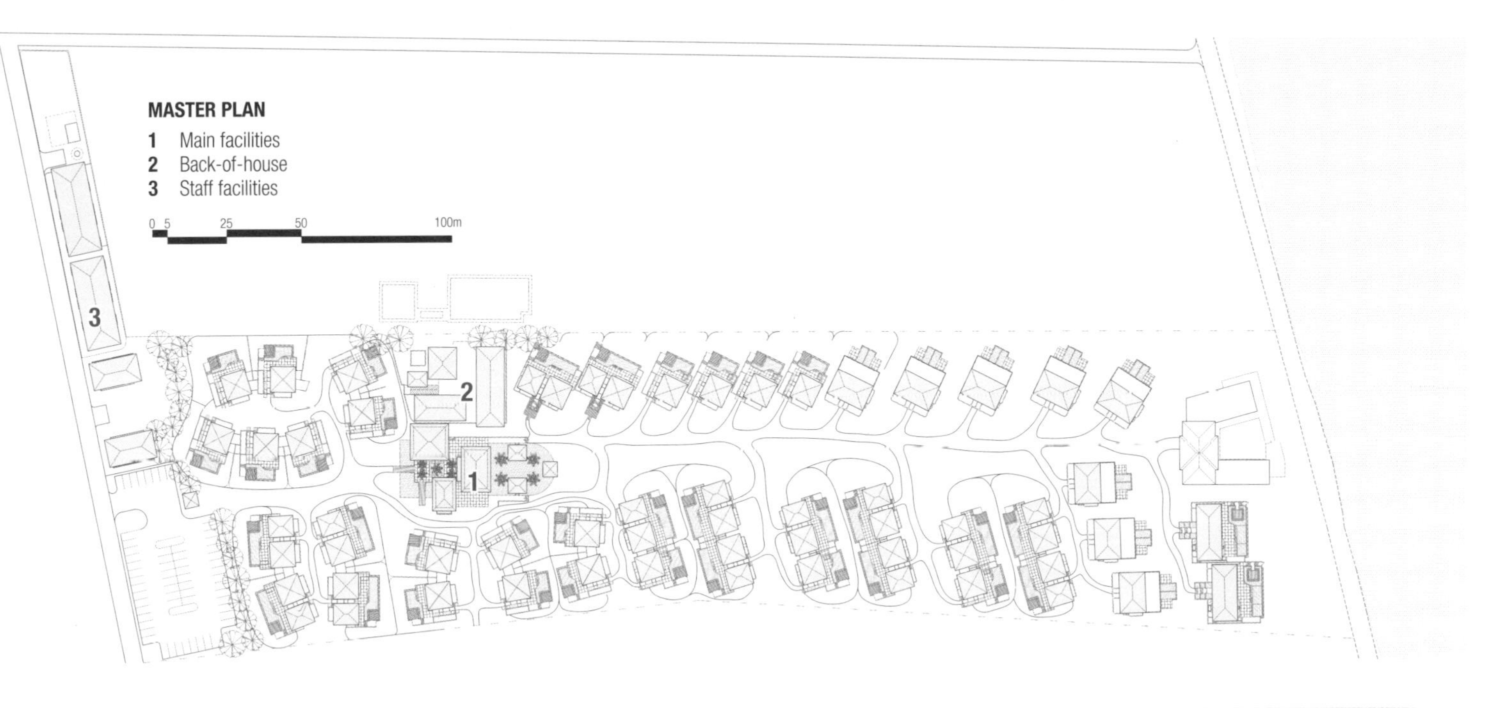
MASTER PLAN
1 Main facilities
2 Back-of-house
3 Staff facilities
0 5 25 50 100m
3
2
1

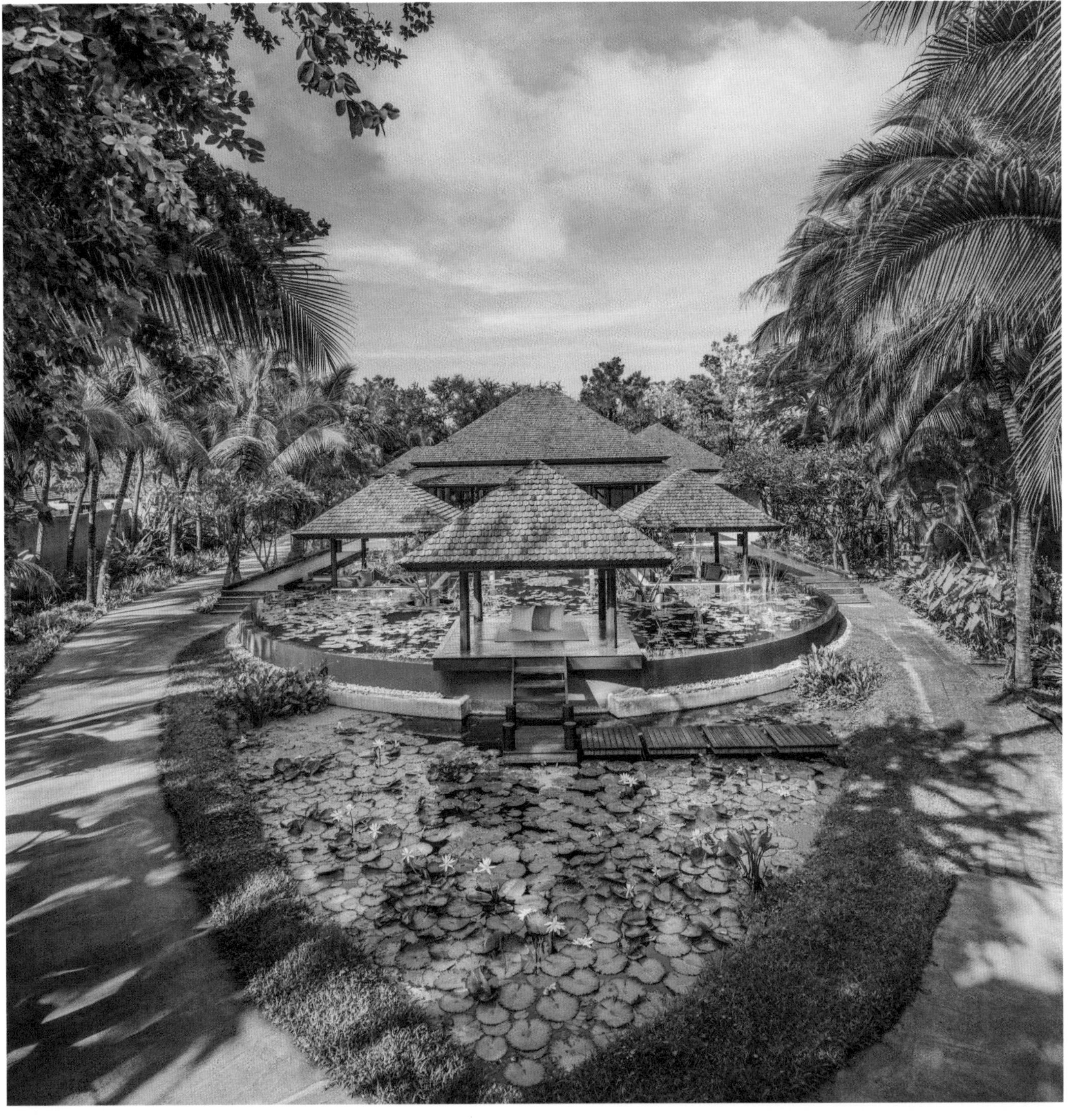

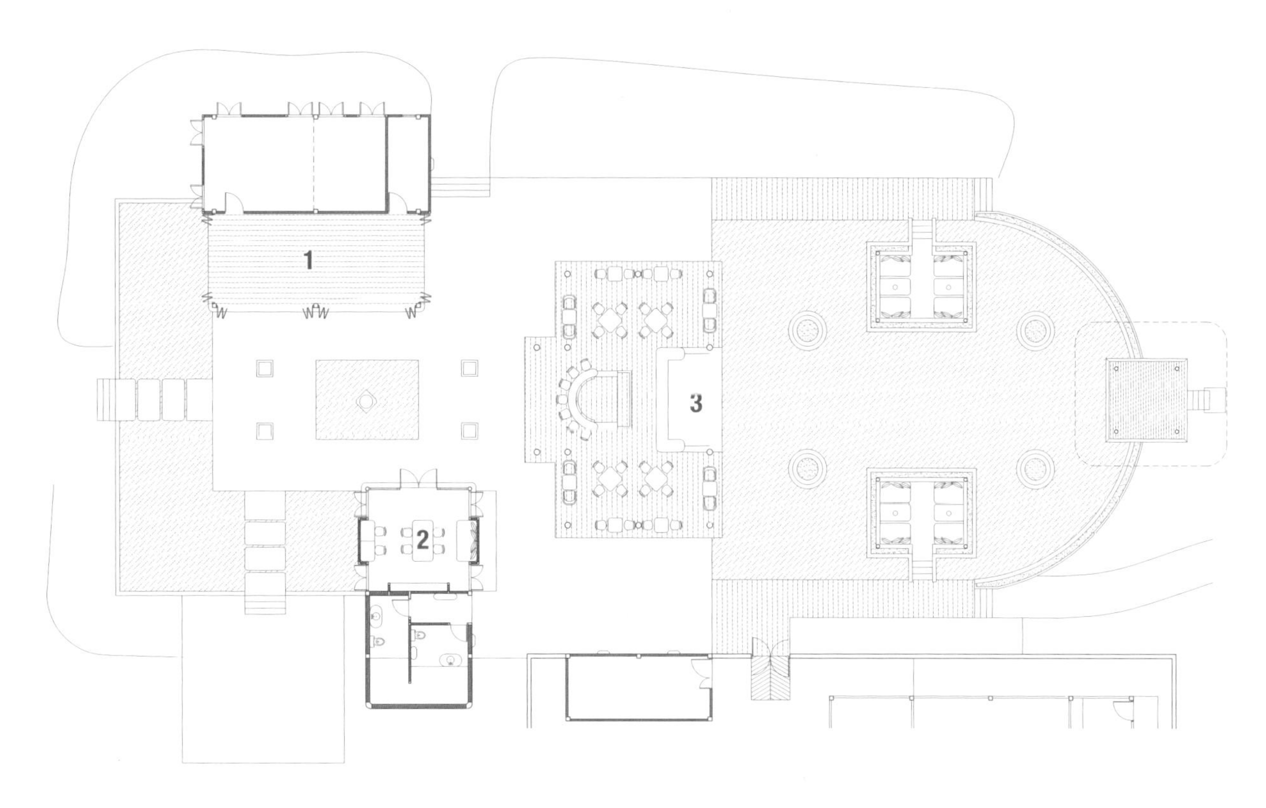

MAIN FACILITIES

1 Reception
2 Library
3 Bar & restaurant

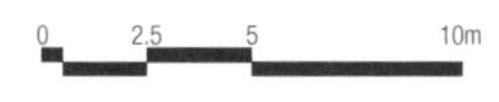

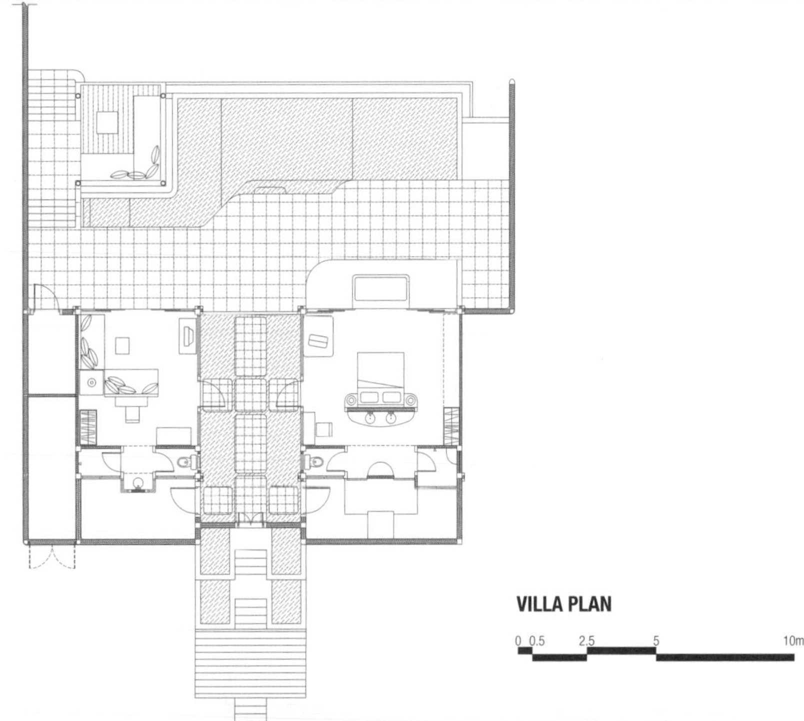

VILLA PLAN

0 0.5 2.5 5 10m

Reception
Link@Sheraton

阁遥岛六善酒店

SIX SENSES YAO NOI

Location Koh Yao Noi, Phang Nga, Thailand
Date 2008
Interior designer Six Senses Creative Department
Landscape designer Inside Out Design

Koh Yao Noi is an unspoilt island located in the Andaman Sea halfway between Phuket and Krabi. Our initial site visit left a long-lasting impression on us. Approaching the island by speedboat presented impressive views of Pah Koh, the Thai name for the fabulous, lop-sided limestone karst islets for which the area is famous.

Upon landing on the site, thickly wooded with lush rainforest, we met a family of hornbills, their throats and bills splendidly colored.

These initial experiences of the site's natural wealth inspired us, sparking a determination to touch this place as lightly as possible with the proposed resort development. The architectural components were derived from local vernacular housing—the hipped roofs typical of southern Thai architecture. We wanted these roofs, as the dominant manmade elements on the site, to age naturally and gradually amalgamate with the jungle. So, local sugar palm leaf was selected as the roofing material. The thatch is retrained by timber lattice at the ridges—adapting an age-old local method for enduring monsoonal storms.

Exhaustive efforts were made to sub-divide the buildings—all 56 pool villas and public buildings—into small components and to insert them onto the steep topography of the site. There was minimal disturbance to the existing ground and to existing trees. The extensive covered outdoor living areas throughout—such as restaurants and transitional verandas between bedroom and pool in the villas—lend transparency to the structures, again contributing to the objective of the resort very gently impacting the site.

We are gratified to hear feedback from visitors to Yao Noi saying that it's as if the buildings have been there for a long time. This is partly due to the way the materials have weathered. It's also a result of simplifying the structure and honing architectural components to their essence. Views to the timeless, otherworldly Koh Pah are framed by columns hewn from timber logs, thatch roof, and infinity edge pool. Even more gratifying is that a family of hornbills have returned to nest here. We like to think this is the same family we met on our first visit.

MASTER PLAN
1 Entrance
2 Guest relations
3 Bar
4 Shop
5 Kitchen
6 Dining pavilion
7 Spa
8 Pool Villa
9 Pool Villa Suite
10 Deluxe Pool Villa
11 The Retreat
12 The Reserve
0 5 25 50 100m
N

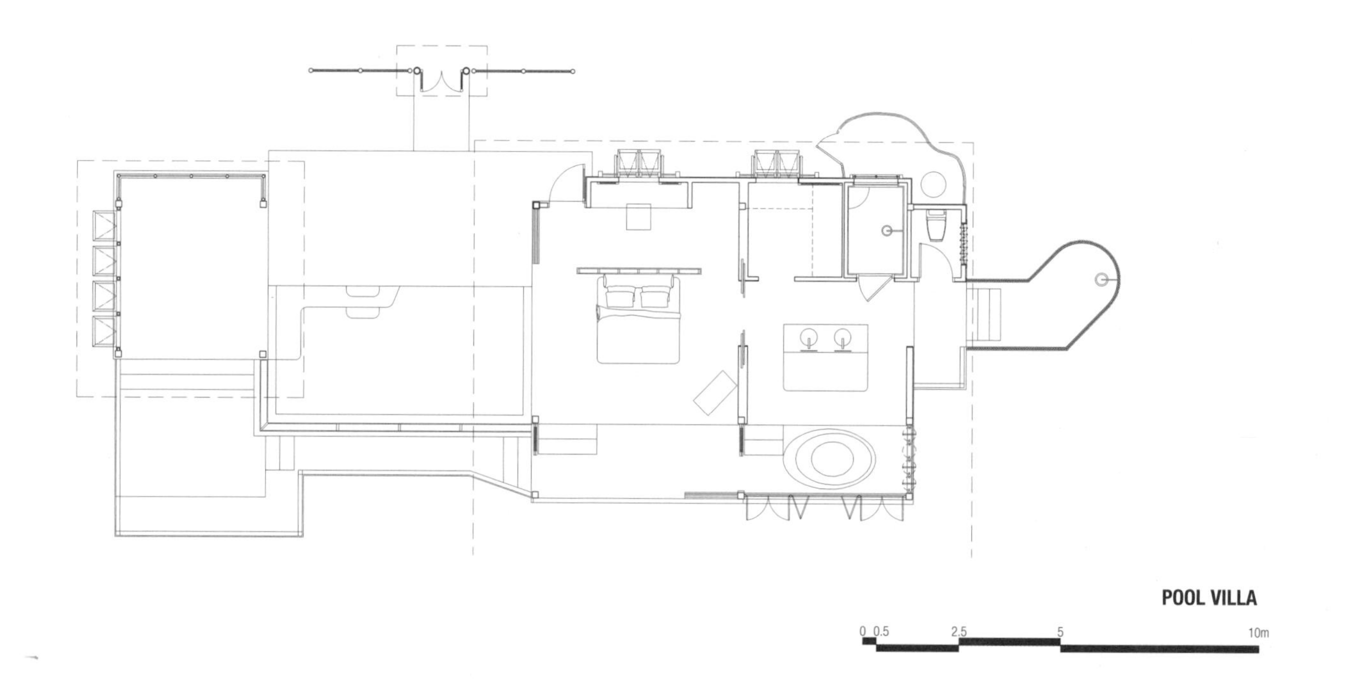

POOL VILLA

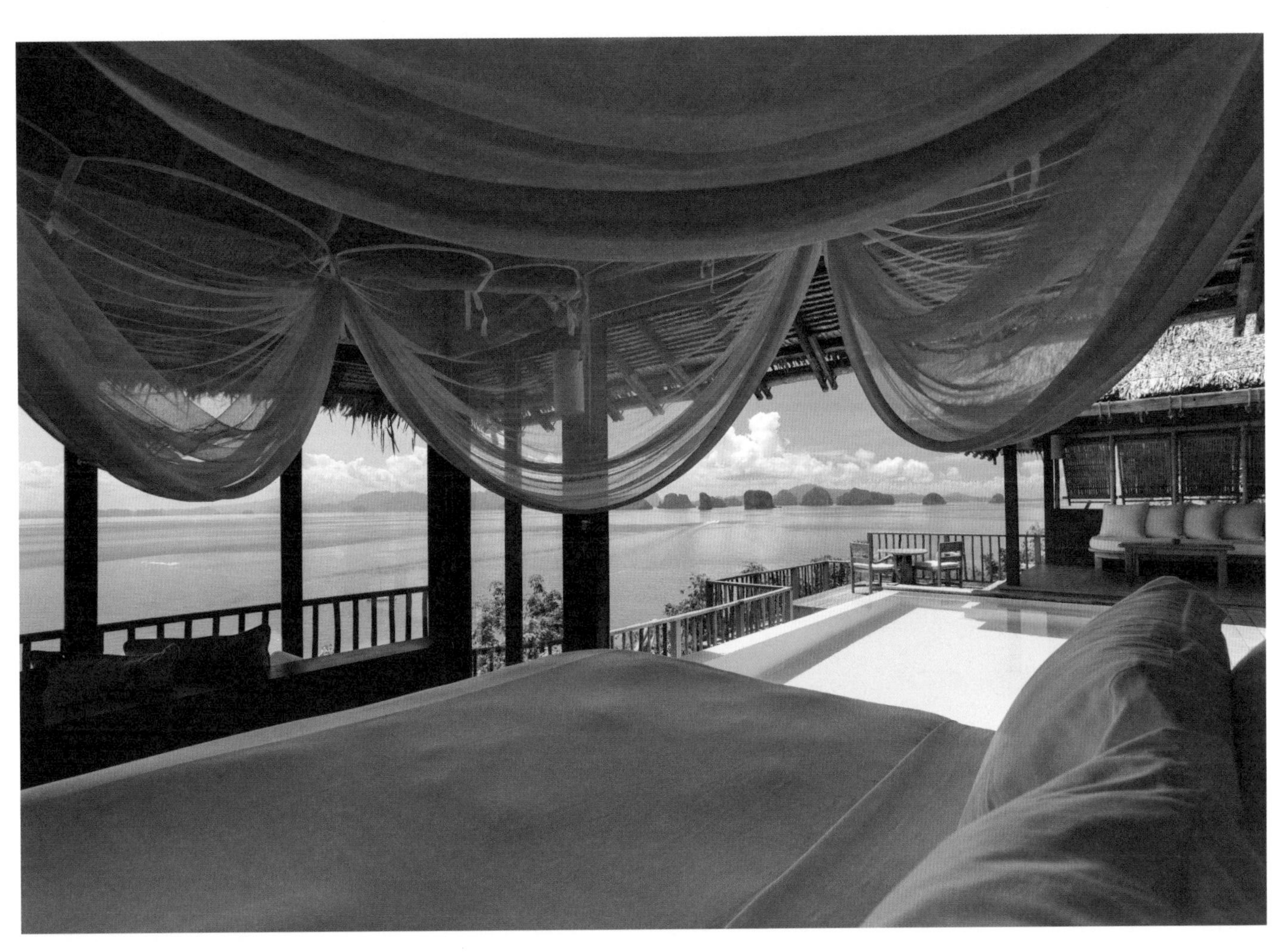

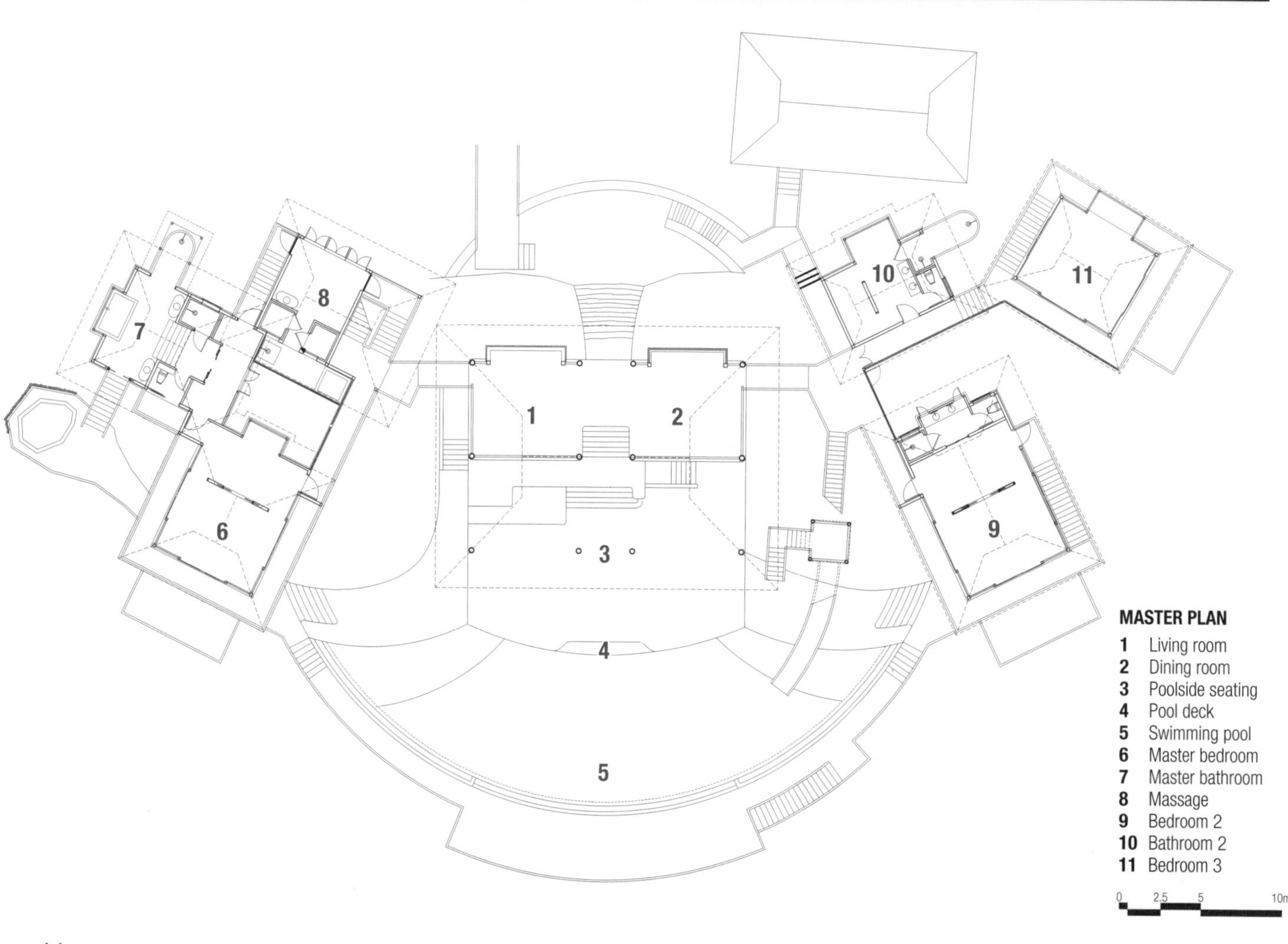

MASTER PLAN

1 Living room
2 Dining room
3 Poolside seating
4 Pool deck
5 Swimming pool
6 Master bedroom
7 Master bathroom
8 Massage
9 Bedroom 2
10 Bathroom 2
11 Bedroom 3

普吉岛小长岛六善酒店

SIX SENSES DESTINATION SPA PHUKET

Location Koh Naka Yai, Phuket, Thailand
Date 2008
Interior designer Six Senses Creative Department
Landscape designer P Landscape

The principal design objective of Six Senses Destination Spa Phuket (now known as the Naka Island, A Luxury Collection Resort & Spa Phuket) was to create a retreat resort focused on guest wellness. It also had respect the existing environment and set an example for sustainable tourism. To deliver this objective, the team created an integrated architectural and landscape concept of sustainable living. Buildings are extremely sensitively designed, both in scale with their surroundings, while the guest experience closely interacts with nature. A number of ecological and biodiversity strategies were incorporated into our design. These included extensive plantings of 'edible gardens' and the use of arboriculture techniques designed to ensure the healthy growth of the existing trees. Large trees are also incorporated as sun shade and visual landmarks in the overall landscape design.

In focusing on preservation of the existing ecosystem in this way, we hoped the guest experience would be one of harmonization between the environment and their daily activities.

The project is located on Naka Yai Island, a small island near Phuket. The only access to the resort is by boat. This allows guests time to unwind from the clamor of Phuket town during their journey across to the island, in preparation for a place of true retreat for mind and body. The jetty faces the Phuket mainland, sheltered from wind and waves in the rainy season.

While the island is generally hilly, a large piece of flat beach area makes the perfect location for the guest welcome area, and the spa, main restaurant, and pool areas.

The priority given to wellness is evident in the proportion of the site and built area given over to the spa facilities. Occupying about 20 percent of the site, the area consists of four separate treatment areas, each inspired by a nationality famed for their particular therapeutic practices:

Indian concept	*Curative*
Chinese concept	*Preventative*
Indonesian concept	*Beauty / relaxation*
Thai concept	*Equilibrium*

All 67 pool villas are located to maximize their views to the sea. Fortunately, the site is a peninsula with a hill running down its spine. This topography provides for villa sites on both sides of the ridge, all with unobstructed sea views and minimal impact on the existing ecology.

MASTER PLAN
1 Back-of-house
2 Main facilities
3 Main swimming pool
4 Spa area
5 Function room
6 Cluster Villa
7 Beach Pool Villa
8 Hill Top Pool Villa
9 Presidential Villa
10 Juice bar
0 5 25 50 100m

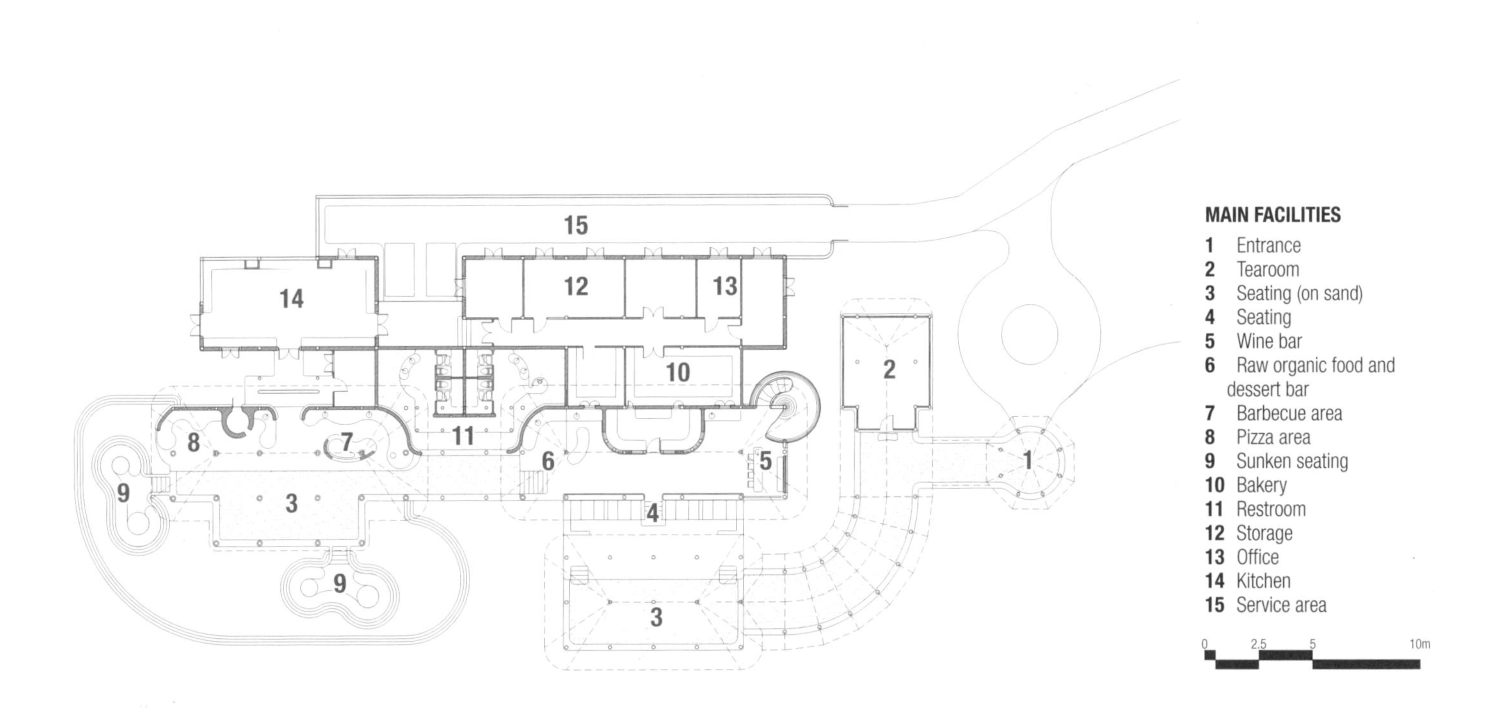
MAIN FACILITIES
1 Entrance
2 Tearoom
3 Seating (on sand)
4 Seating
5 Wine bar
6 Raw organic food and dessert bar
7 Barbecue area
8 Pizza area
9 Sunken seating
10 Bakery
11 Restroom
12 Storage
13 Office
14 Kitchen
15 Service area
0 2.5 5 10m

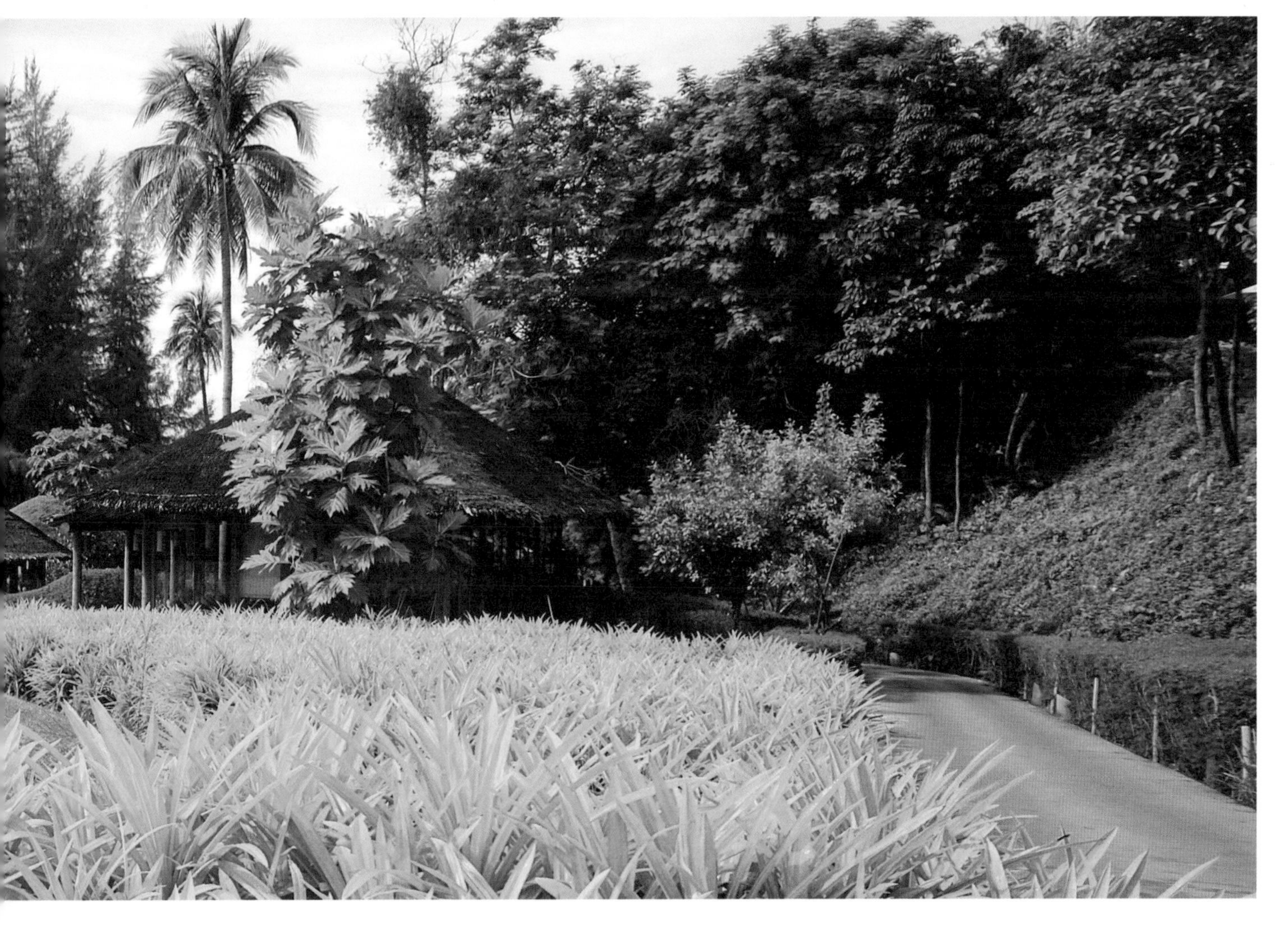

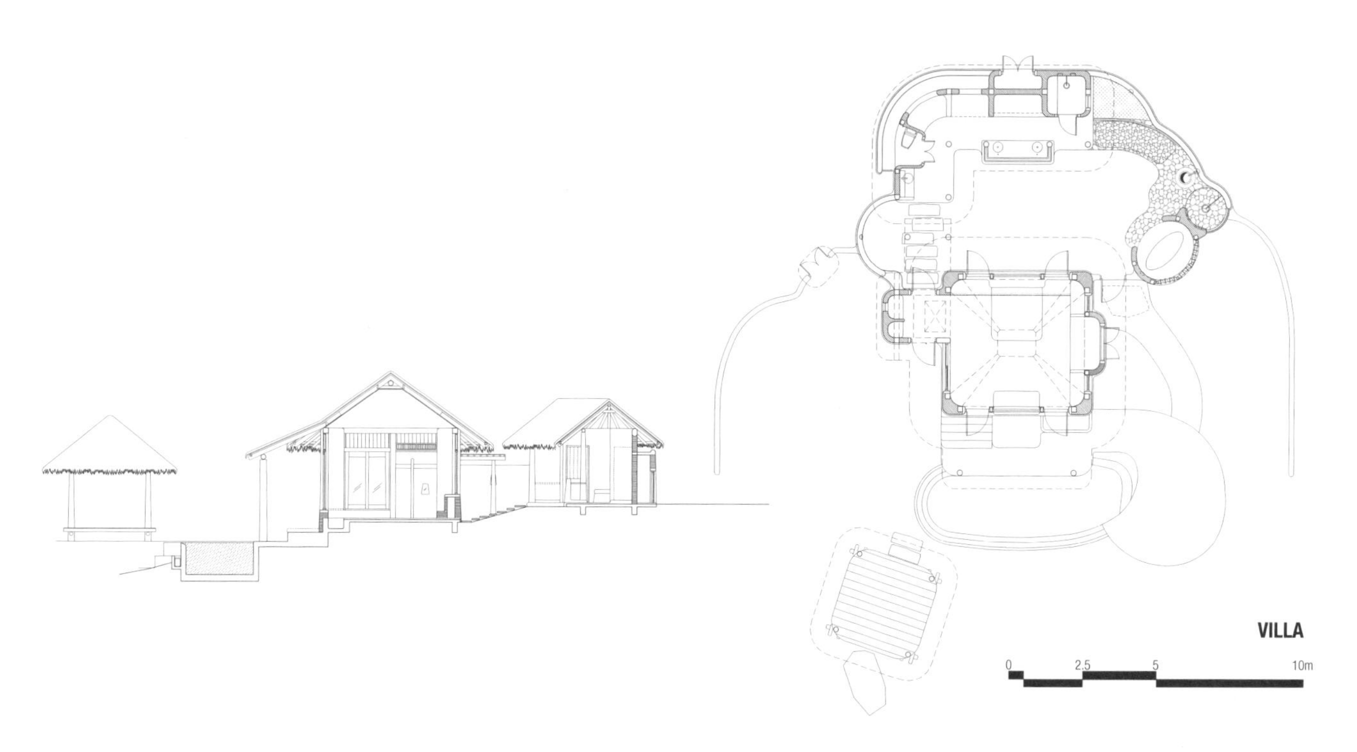
VILLA
0
2.5
5
10m

拉姆岛六善酒店

SIX SENSES LAAMU

Location Olhuveli, Maldives
Date 2011
Interior designer P49 Deesign & Associates

The topographical signature of the Maldives is flatness. The 26 atolls are formed on the semi-submerged rim of ancient volcanoes, with the average height of land above sea level only 5 feet (1.5 meters).

This horizontality is evident in the overall resort layout of Six Senses Laamu, in which buildings are dispersed evenly on land and over the water. They appear to first inhabit the island, located at the southwestern tip of the atoll, and then be in the process of spreading out judiciously to the northeast, over the turquoise lagoon of Laamu Atoll.

Being the only resort on this atoll, the resort's embrace of the water has an exploratory character. The 70 water villas are spread evenly over three jetties in a very geometric and rhythmic order. In contrast, the main cluster of public facilities (including arrival jetty, bars, restaurants, and dive school), which is also built over the water, is crouched precariously close to the western edge of the atoll. Just beyond, the water color changes abruptly from safe turquoise to unknown indigo blue.

On the island, guest facilities include 26 beach villas, wellness facilities, and garden restaurant. Back-of-house facilities and staff accommodations are concealed in the center of the island.

The exploratory nature of the guest experience is exhibited throughout the resort by the buildings' interaction with nature.

For example, the four couple's treatment nests that comprise the spa are directly inspired by the tangled birds' nests discovered atop the screw pine forests that cover the island. At the same time, the random pattern of woven branches that make up the semi-transparent external shell of these nests provides only a modicum of privacy. Guests are barely shielded from the sights and sounds of the forest outside.

Interaction with nature is also encouraged by the provision of plentiful catamaran nets suspended out above the water in the public facilities. The visitor can even retire directly onto such a platform, positioned right beside specially placed dining tables. In the water villas, opportunities for interaction with the lagoon abound, from open-air showers, glass-over-water bathtubs, and the option to set off snorkeling directly from the veranda of each guestroom.

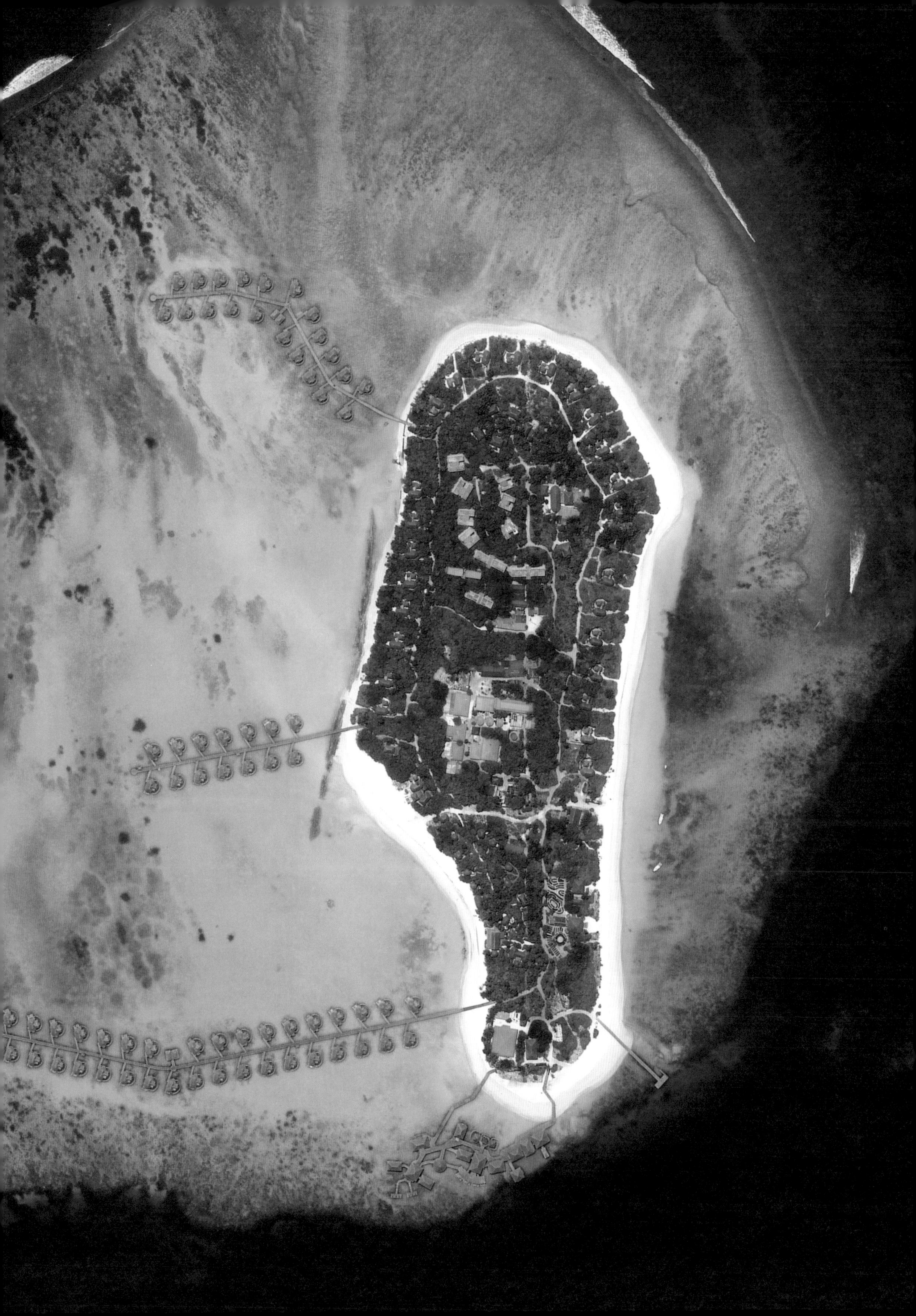

MAIN FACILITIES
1 Welcome sala
2 Guest relations / library
3 Shop
4 Dive center
5 Ice-cream parlor
6 Bar
7 Lounge
8 Guest's WC
9 Cheese & wine cellar
10 Dining pavilion
11 Cooking hut
12 Kitchen
13 Restaurant
14 Dining platform
0
5
25
50
100m
N

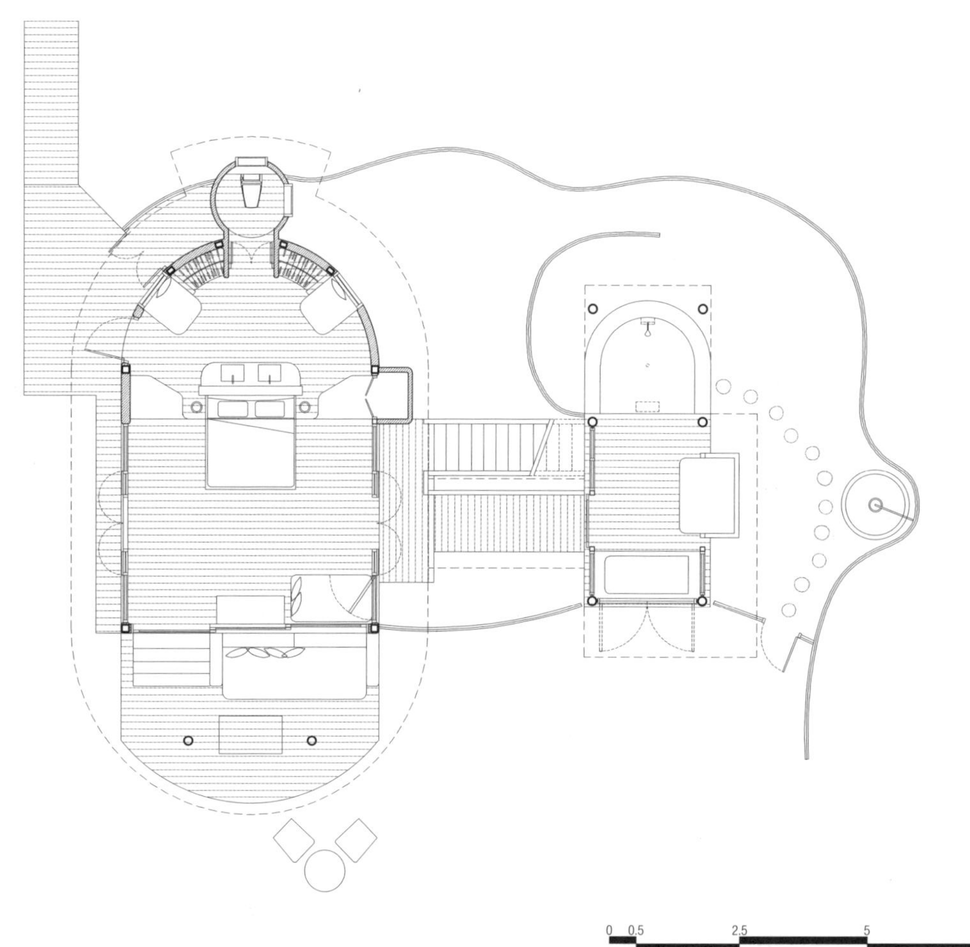

BEACH VILLA

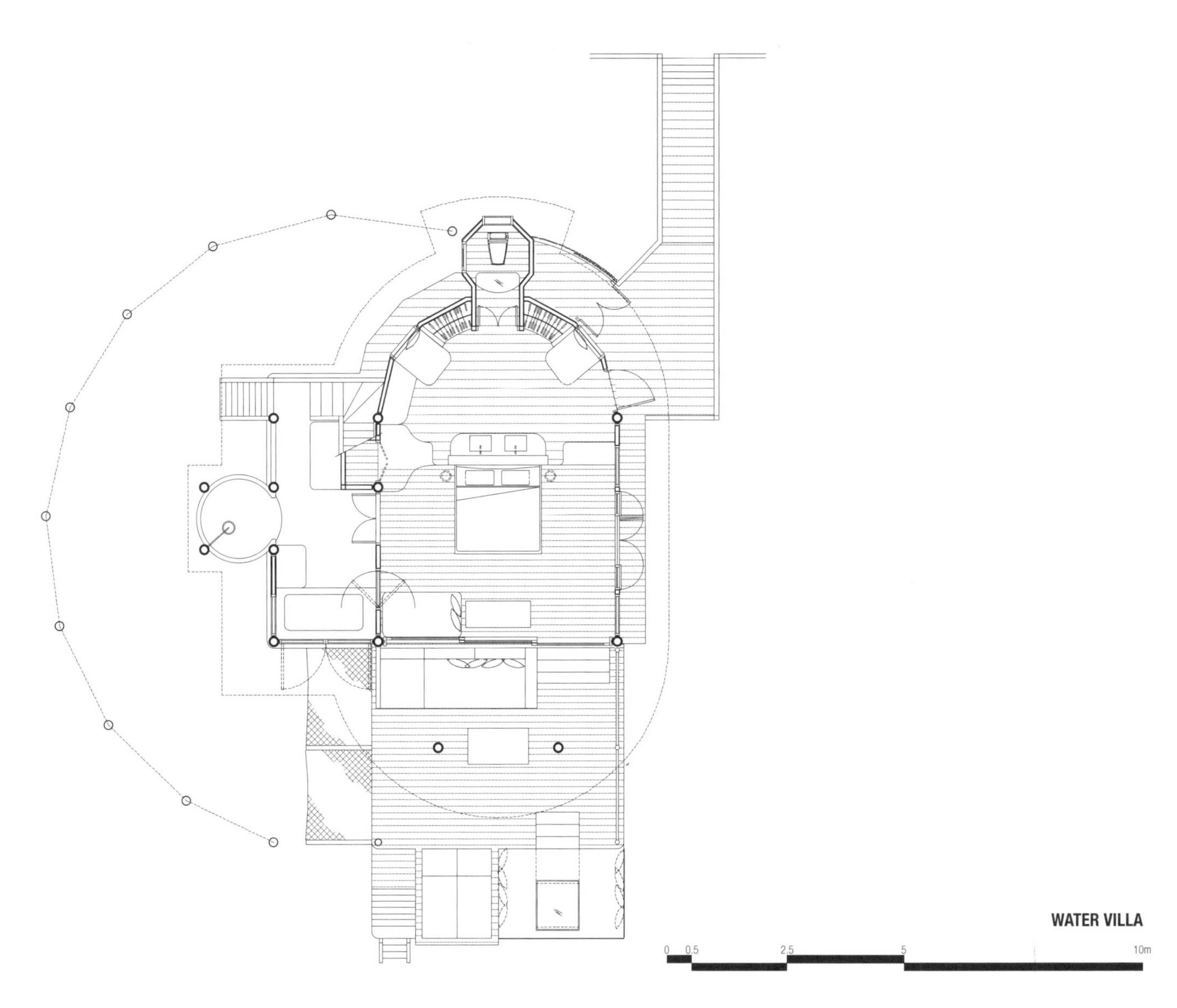
WATER VILLA
0
0.5
2.5
5
10m

青城山六善酒店

SIX SENSES QING CHENG MOUNTAIN

Location Chengdu, China
Date 2015
Interior designer Six Senses Creative Department
Landscape designer AECOM Hong Kong

This project, situated in a transcendental landscape in Sichuan province, was an opportunity for Habita to articulate our respect for traditional Chinese garden architecture.

This resort focuses on Mount Qing Cheng, one of the most important Taoist centers in China. Starting with a respectful gaze to the mountain, Feng Shui principles determine the layout of the site—main and secondary entrances, and the flow of water throughout the site.

Principal buildings are imbued with their own characteristics, but of paramount importance are harmonious visual connections between mountain, water, nature, and other structures.

The Welcome Pavilion epitomizes the connection between nature and the resort buildings. It is conceived as a bridge, with the main hall suspended between two mountains. Guests arrive at the resort from a dense bamboo avenue leading to the top of the mountain. Entry via a massive antique door reveals a panoramic view over the tiled resort roofscape. It is as though the visitor has arrived at a secret village, humbly located at the foot of Mount Qing Cheng.

The main restaurant also exhibits its own architectural vocabulary, its strong circular form over three floor levels alluding to notions of infinity. But this visually compelling building is also subservient to its axial relationship between Welcome Pavilion and mountain.

Other main buildings are integrated more subtly with the landscape. External pool and spa blend with mini-island gardens, entered via tunnels. The landscaping design itself relies on functionality to convey a sense of humility and authentic village experience. Duck ponds integrated with rice paddies and market gardens feature prominently in the landscaping scheme.

Arranged in small clusters and rows, 113 rooms and villas complete this pastoral scene. External materials comprise clay roof tiles, natural plaster, and natural timber gable ends, which are derived from a traditional timber structure. Each unit is carefully oriented to provide guests with both an unrestricted view to the mountain and their own privacy, despite space being very limited. The interiors are contemporary. General details are derived from the Chinese courtyard house, with the exception of modern light-colored finishing materials, which are utilized throughout, completing the guest experience of tranquility and humble appreciation of the location.

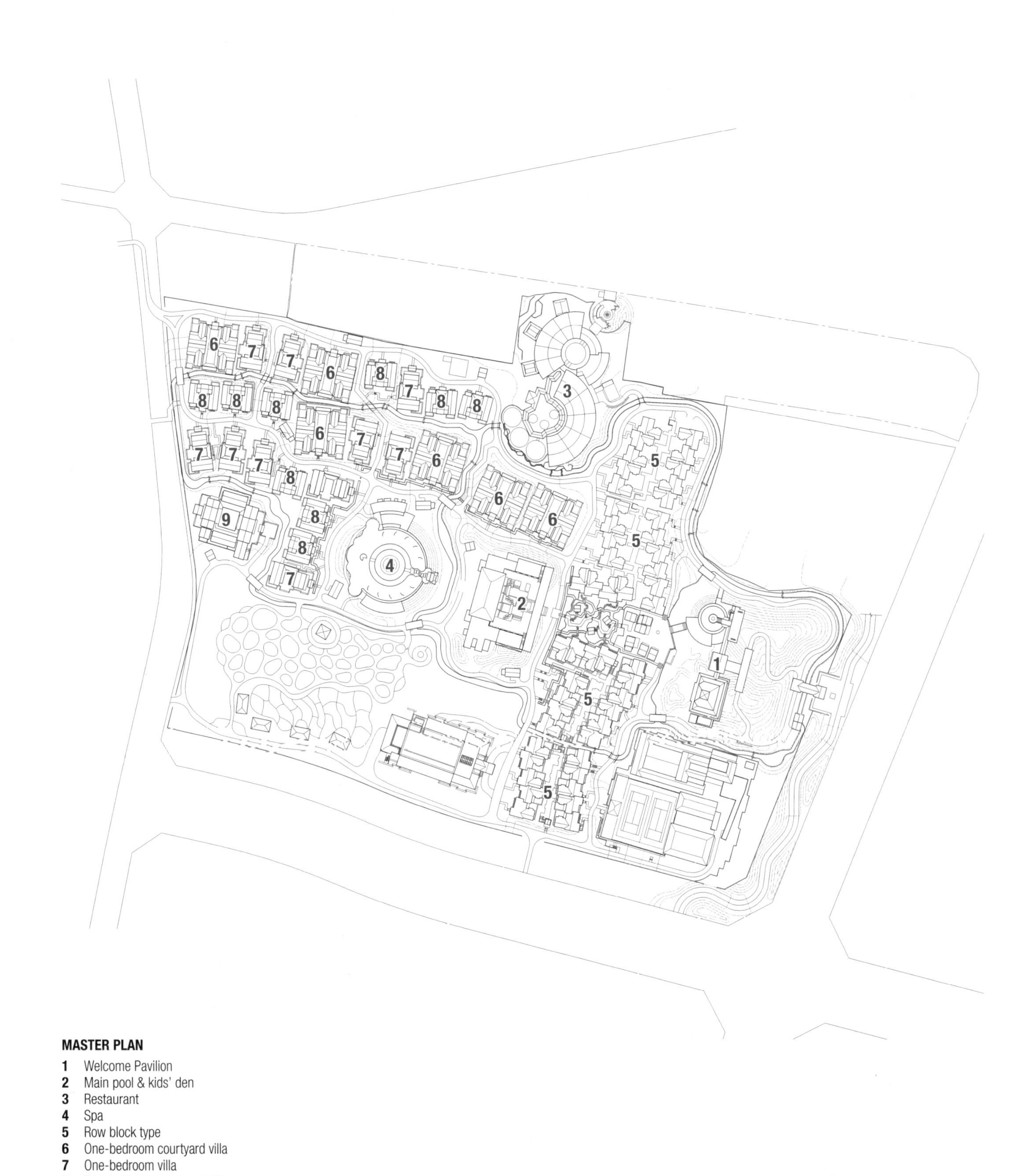

MASTER PLAN

1 Welcome Pavilion
2 Main pool & kids' den
3 Restaurant
4 Spa
5 Row block type
6 One-bedroom courtyard villa
7 One-bedroom villa
8 Two-bedroom courtyard villa
9 Owner villa

0 5 25 50 100m

CHINESE PRIVATE DINING

1 Entrance
2 Reception
3 Dining
4 Boardroom
5 Restroom

0 1 5 10 20m

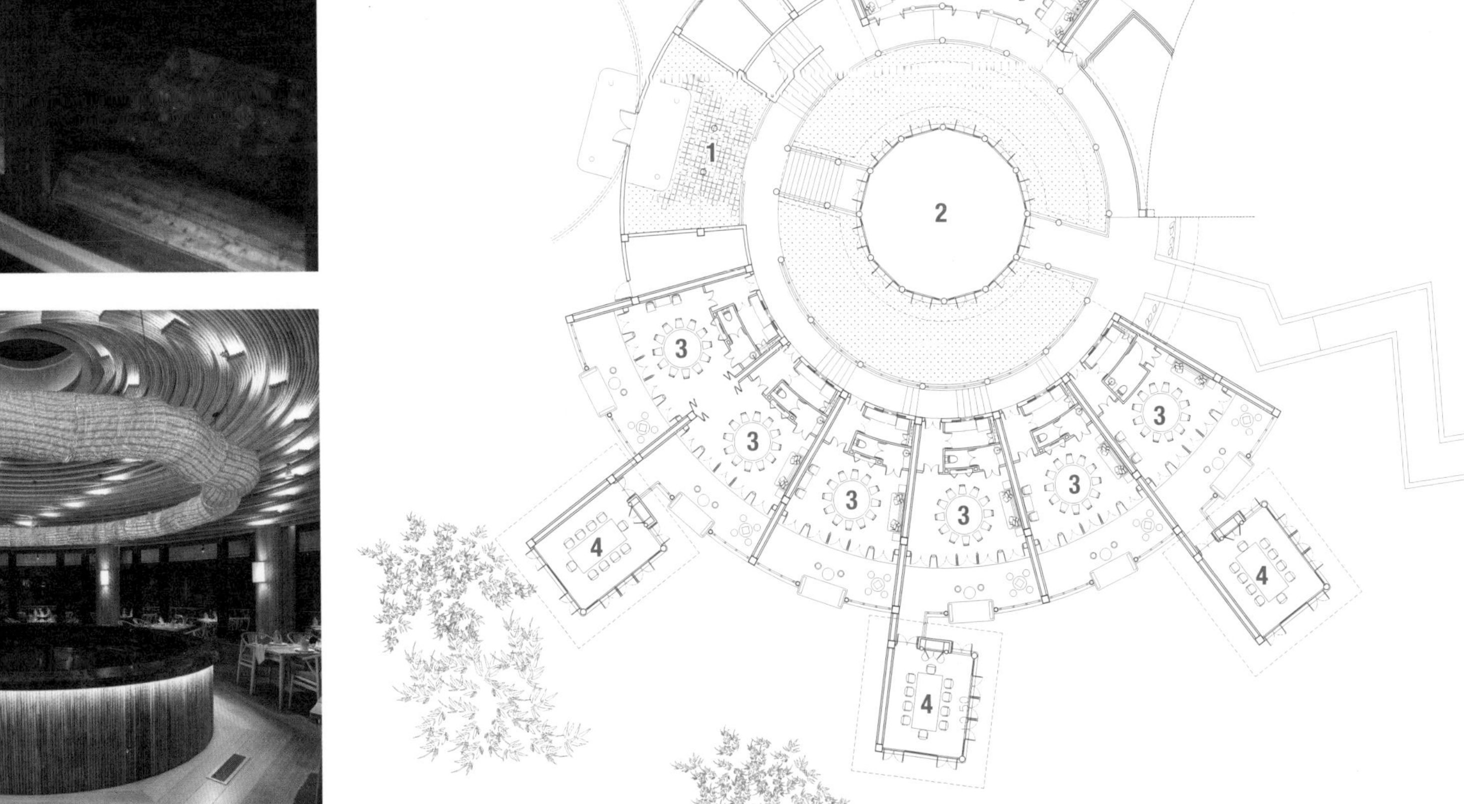

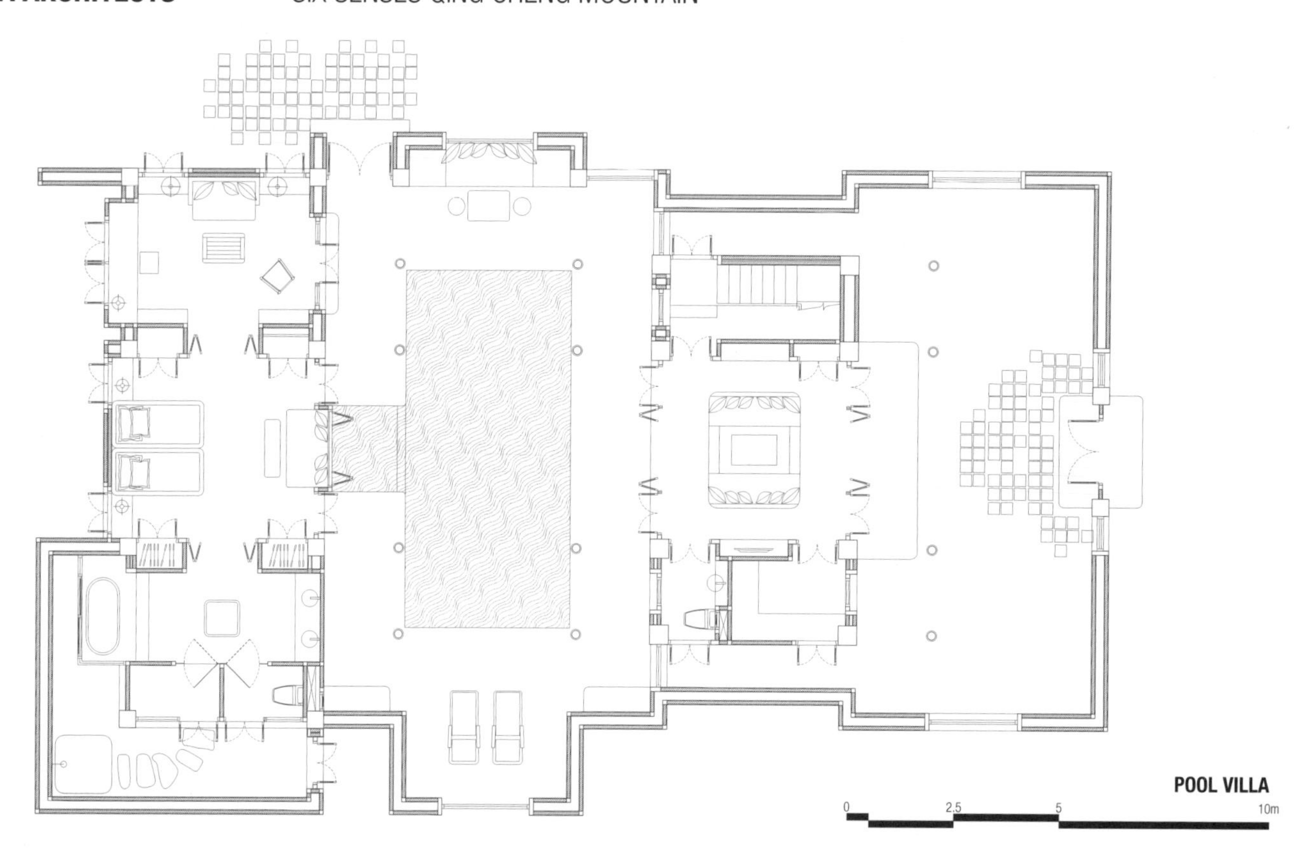
POOL VILLA
0
2.5
5
10m

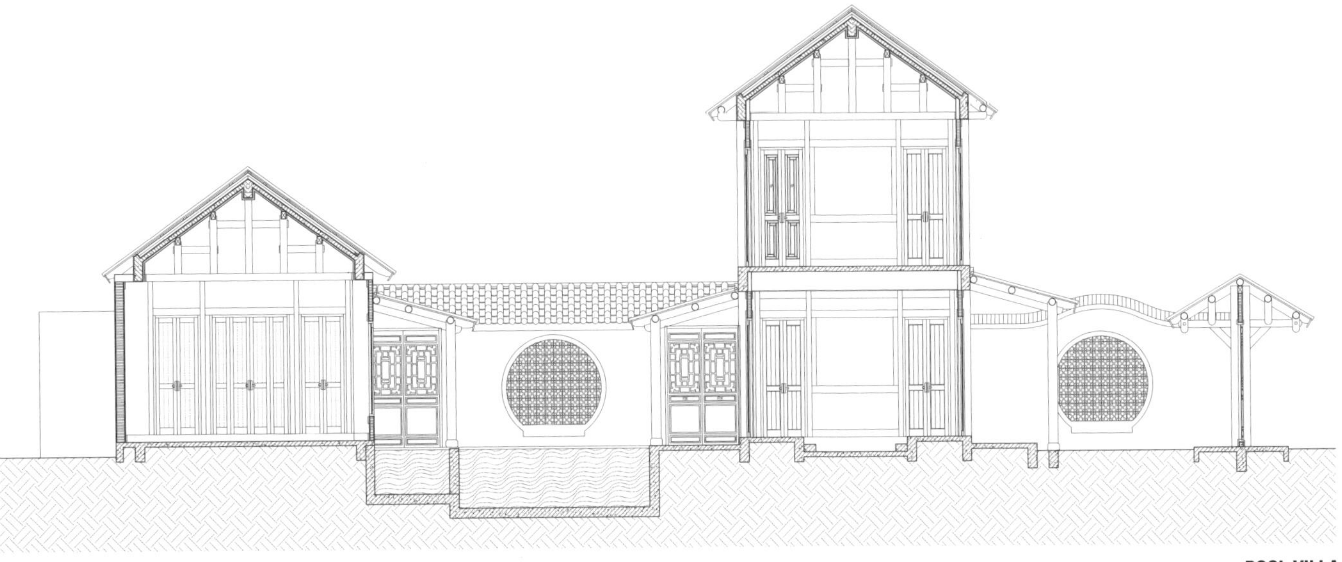

POOL VILLA

索尼娃奇瑞度假村
SONEVA KIRI

Location Koh Kood, Trat, Thailand
Date 2009
Interior designer Soneva Creative Department

Soneva Kiri is an exclusive resort located on Koh Kood, an island in the Gulf of Thailand. The resort successfully fuses exclusivity with an overall casual atmosphere. This informality is reflected in the architecture, comprising two simple vocabularies that address the beach or hillside aspects of the site.

The beach-oriented buildings—beach villas, main restaurant and pool—feature large white overhanging tent roofs, emblematic of Soneva Kiri when viewed from the sea and air.

For buildings on the hillside sites, traditional pitched roofs are used. Owing to Koh Kood's high rainfall, the roof angles are steep. The hillside villas are set back inconspicuously from the resort's access roads. In the manner of traditional Thai houses, they are designed as a building compound around a timber deck. This allows natural light and wind to circulate and large existing trees to remain undisturbed.

There are 46 villas overall, comprising one-, two-, three-, four-, five-, and six-bedroom types. Natural materials prevail—plantation eucalyptus and pine for structure, bamboo for structure and cladding, and earth added to interior renders.

The main fine dining restaurant comprises cantilevered dining platforms overlooking the sea, anchored visually by a traditional pitched roof over the display kitchen. This roof offers a twist on the traditional, with its bamboo structure spiraling upwards to the apex.

The other main facilities revert to the beach-oriented language of white tent roofs. The all-day dining restaurant, retail, and recreation facilities are arranged along an extensive single ramp over 650 feet (200 meters) long. Such a dominant linear form could adopt a formal character. But here the resort still maintains its laid-back easiness. The ramp takes on amorphous planar shapes as it engages the main facilities, making it appear more organic, while cleverly doubling as a roof for back of house areas.

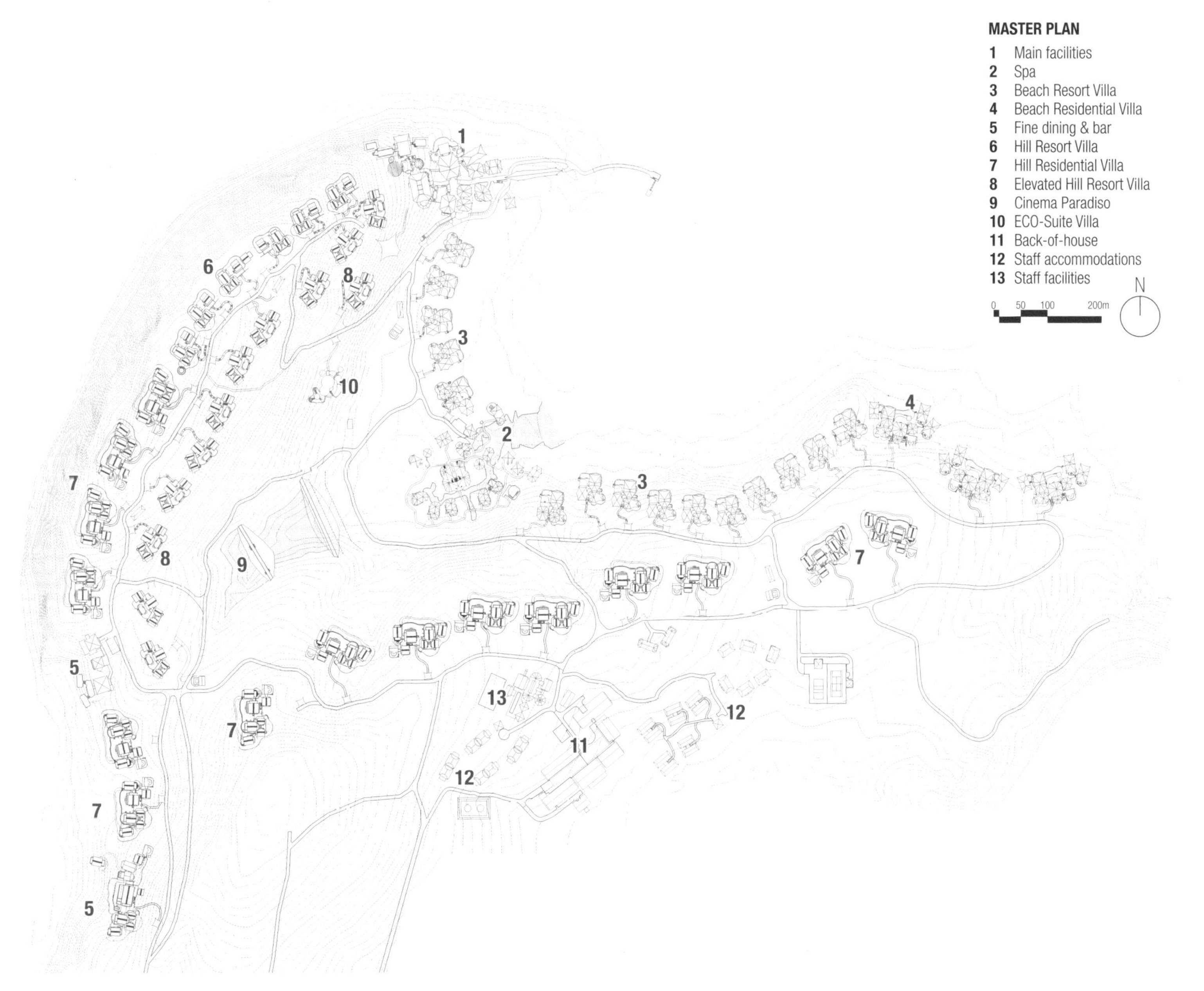
MASTER PLAN
1 Main facilities
2 Spa
3 Beach Resort Villa
4 Beach Residential Villa
5 Fine dining & bar
6 Hill Resort Villa
7 Hill Residential Villa
8 Elevated Hill Resort Villa
9 Cinema Paradiso
10 ECO-Suite Villa
11 Back-of-house
12 Staff accommodations
13 Staff facilities
N
0 50 100 200m

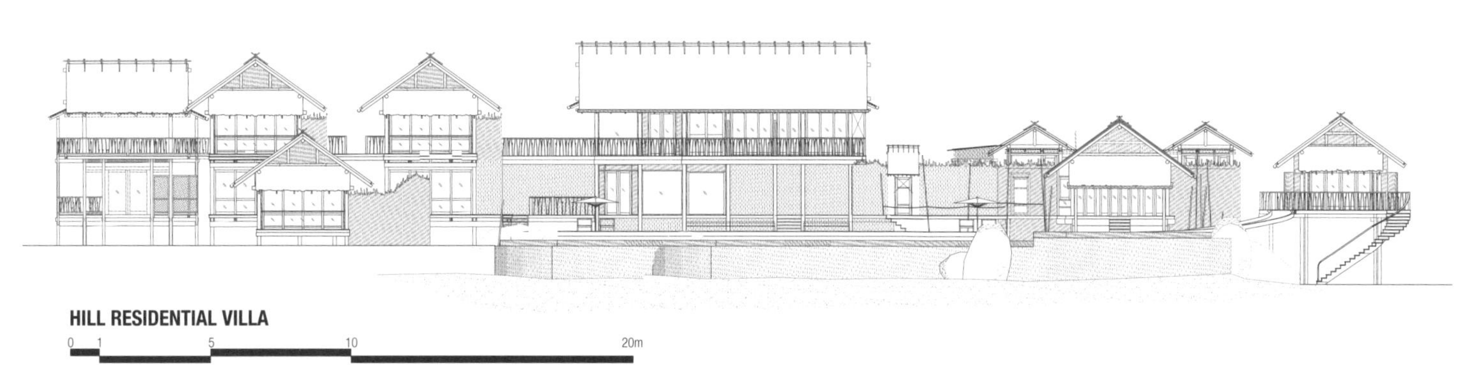

HILL RESIDENTIAL VILLA

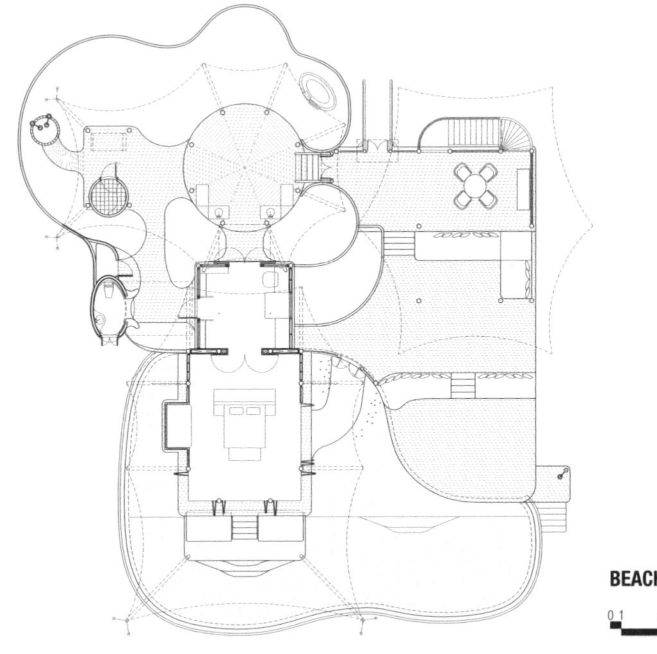

BEACH RESORT VILLA

0 1 5 10 20m

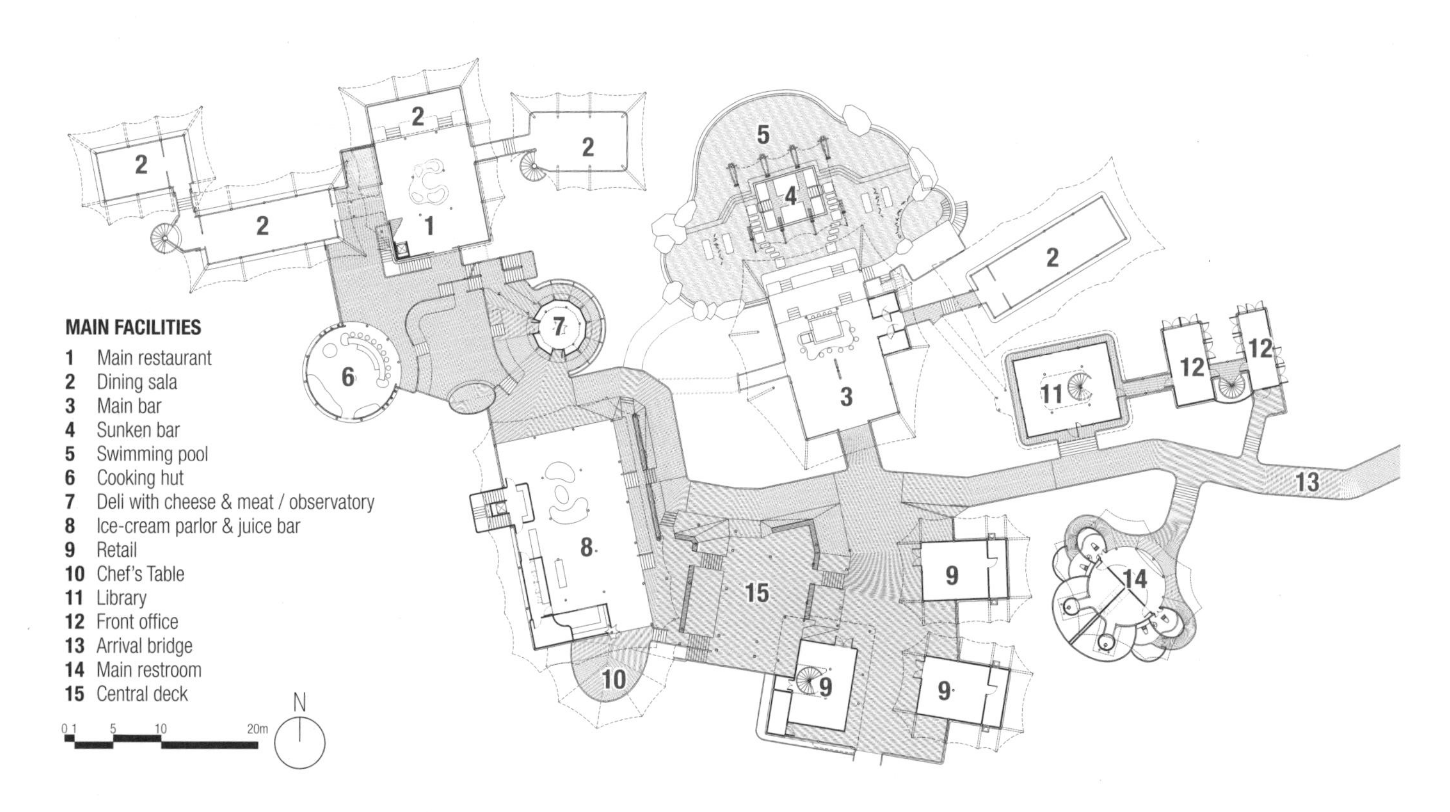

MAIN FACILITIES
1 Main restaurant
2 Dining sala
3 Main bar
4 Sunken bar
5 Swimming pool
6 Cooking hut
7 Deli with cheese & meat / observatory
8 Ice-cream parlor & juice bar
9 Retail
10 Chef's Table
11 Library
12 Front office
13 Arrival bridge
14 Main restroom
15 Central deck
0 1 5 10 20m
N

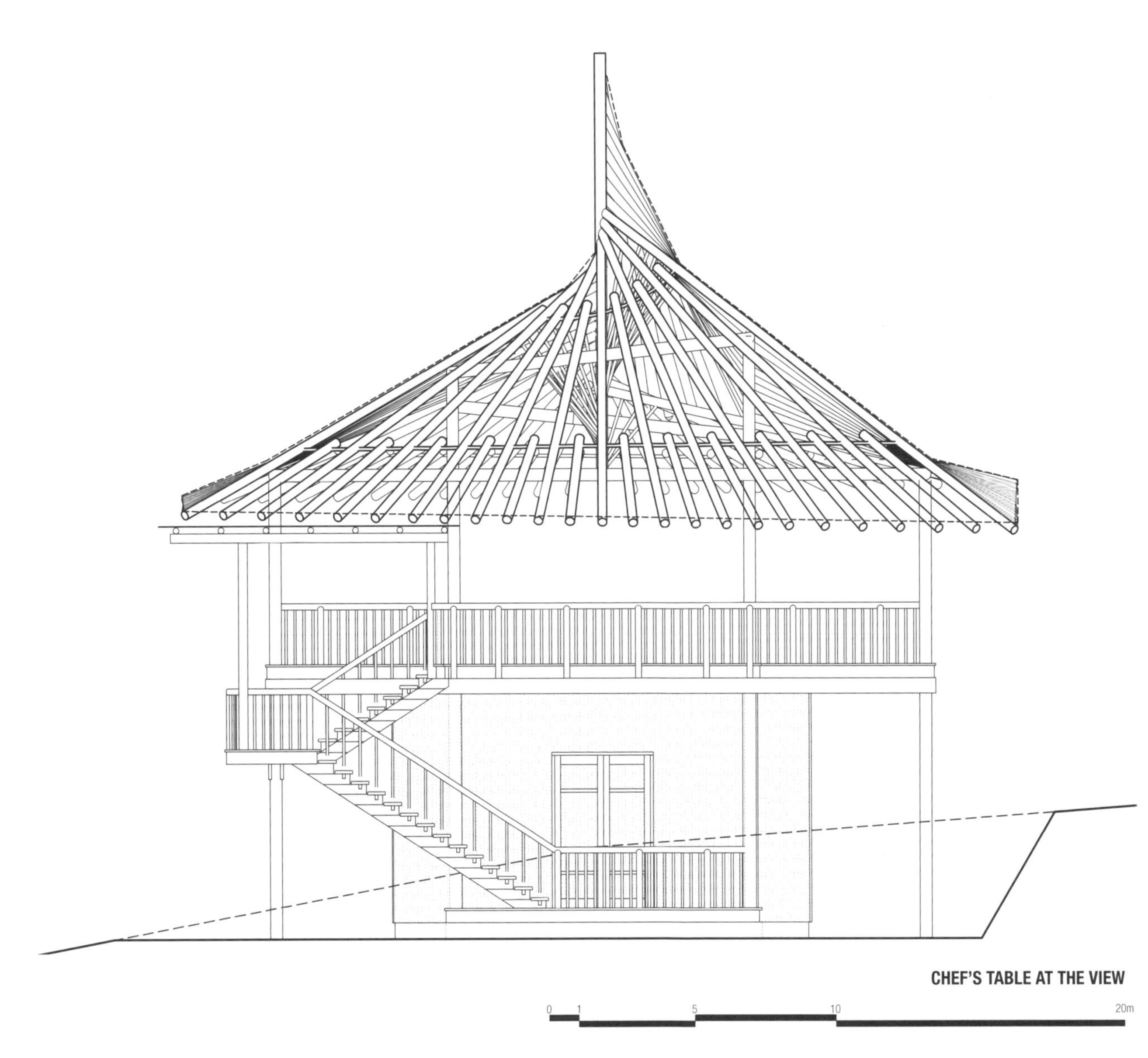

CHEF'S TABLE AT THE VIEW

索尼娃芙西度假村

SONEVA FUSHI

The Crusoe Suite Villa is an additional villa type for Soneva Fushi, developed from the original villa existing on the resort. It differs from the original, mainly by adding more functional area in order to provide more comfortable and convenient guest space.

On the ground floor, a double-volume space was added to the living room. At upper-floor level, this volume reveals a bridge between the bedroom on the upper level and the outdoor roof terrace. This feature gives the Crusoe Suite Villa its unique identity.

The interconnection of upper and lower floors is also provided in the entrance area. As well as connecting the spaces vertically, this double-height arrival space develops a strong connection between indoor and outdoor spaces.

Location: Kunfunadhoo Island, Maldives
Date 2016
Interior designer Soneva Creative Department

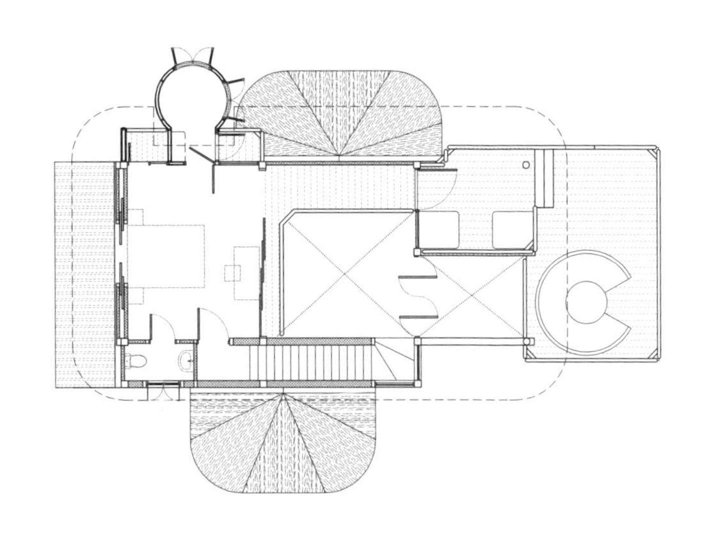

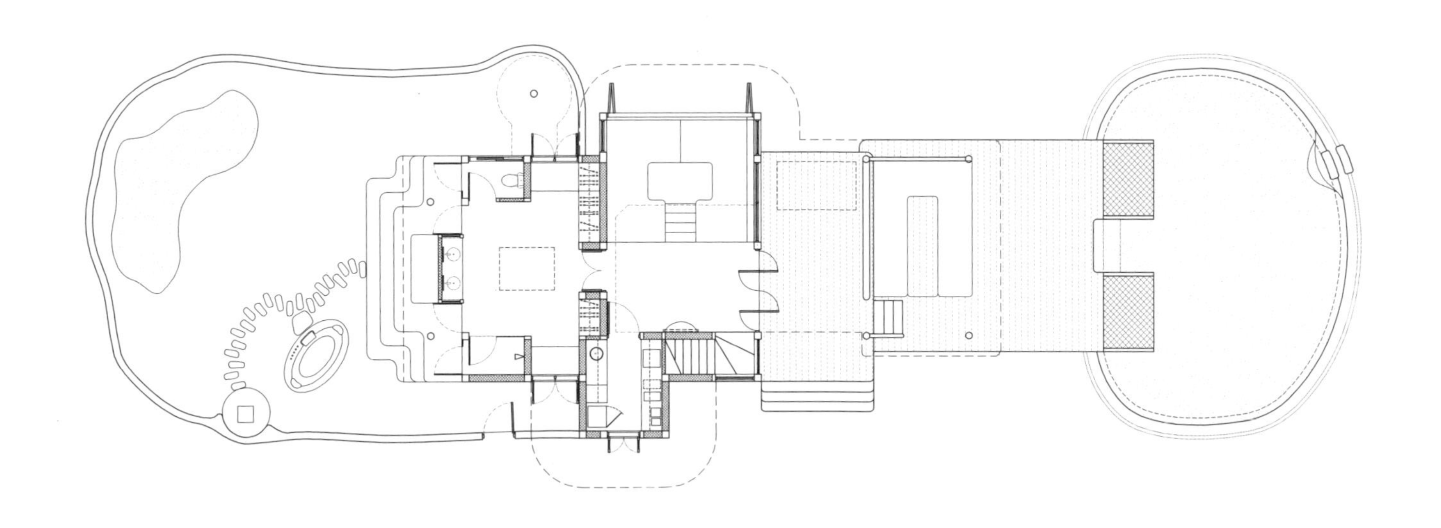

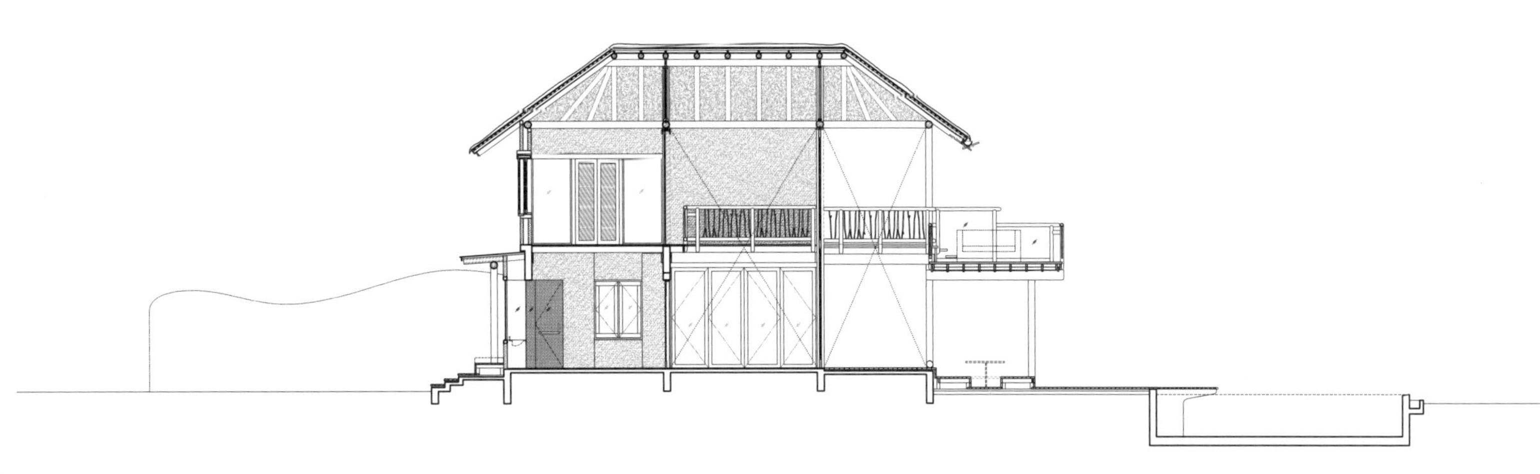

CRUSOE SUITE VILLA

0 2.5 5 10m

索尼娃贾尼酒店
SONEVA JANI

Arriving by sea plane, Soneva Jani first appears to guests like a chain of smoothly rounded driftwood objects, floating out along the tideline north of Medhufaru Island. From directly overhead, the chain of elliptical roofs resembles plankton, geometric yet organic, tethered tenuously to their island host.

At sea level, this floating chain of objects transforms into a luxurious marine ecosystem, perfectly located on Noonu Atoll's 3.5-mile-long (5.6-kilometer-long) crystalline lagoon.

Most of the resort's guest accommodations and facilities are located on this jetty that extends just over half a mile (around 1 kilometer) from the island—24 water villas, restaurants, bars, spa, and diving center. The unifying construction material of this lagoon-bound component of the resort is a sea-bleached timber. Against its intense turquoise background, this material acquires a unique luminosity, yet is natural and appropriate.

The arrival jetty is positioned halfway along the jetty, in front of its largest structure. The three-story The Gathering is the heart of the resort. It houses a variety of dining outlets, as well as the spa, library, and retail area. The Observatory Bar and Marine Biology Centre/Dive School are also clustered in this central location.

To either side of The Gathering is located the guest accommodations, ranging from one-bedroom to four-bedroom water villas. Their elliptical roofs and curving privacy walls shelter living areas and bedrooms, generous sala, and private lagoon plunge pools. All main bedrooms feature a retractable roof that allows stars to observed at night.

Further resort facilities are dispersed across 150-acre (60-hectare) Medhufaru Island. The sparse placement of tennis courts, water sports champa, and Cinema Paradiso restaurant allows the island's lush vegetation to remain intact. This creates shaded forest paths for guests to ramble through, in contrast with the sun-drenched experience of the lagoon. The forest also conceals back-of-house facilities and accommodations for 100 Soneva staff, including a mosque and football field.

Location Medhufaru Island, Noonu Atoll, Maldives
Date 2016
Interior designer Soneva Creative Department

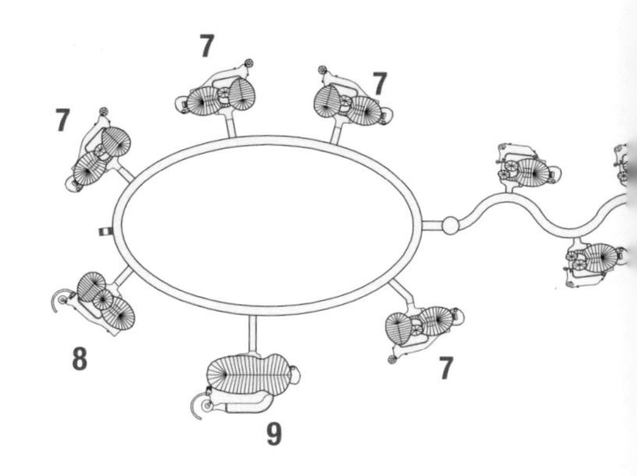

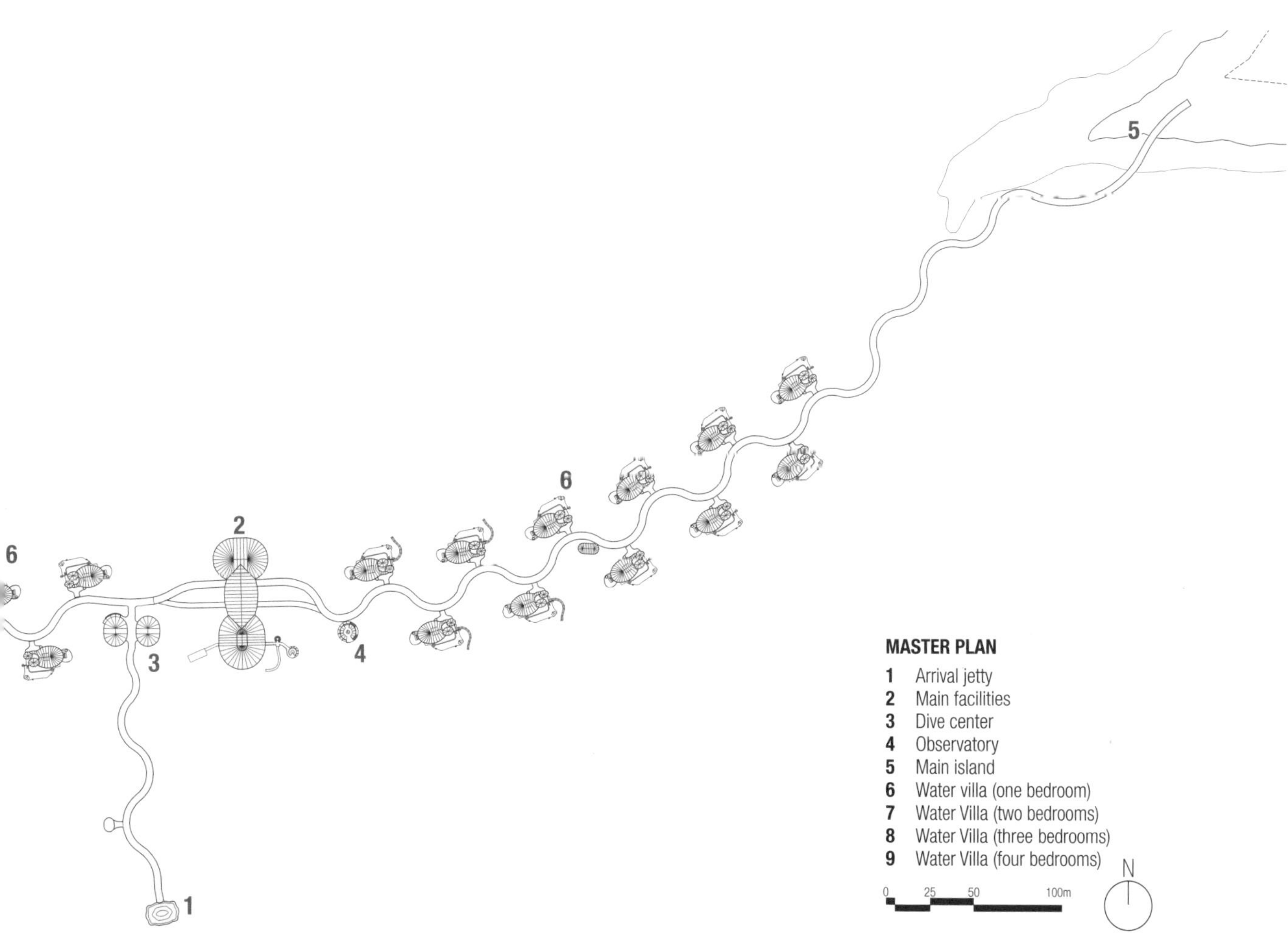

1
2
3
4
5
6
6
MASTER PLAN
1 Arrival jetty
2 Main facilities
3 Dive center
4 Observatory
5 Main island
6 Water villa (one bedroom)
7 Water Villa (two bedrooms)
8 Water Villa (three bedrooms)
9 Water Villa (four bedrooms)
0 25 50 100m
N

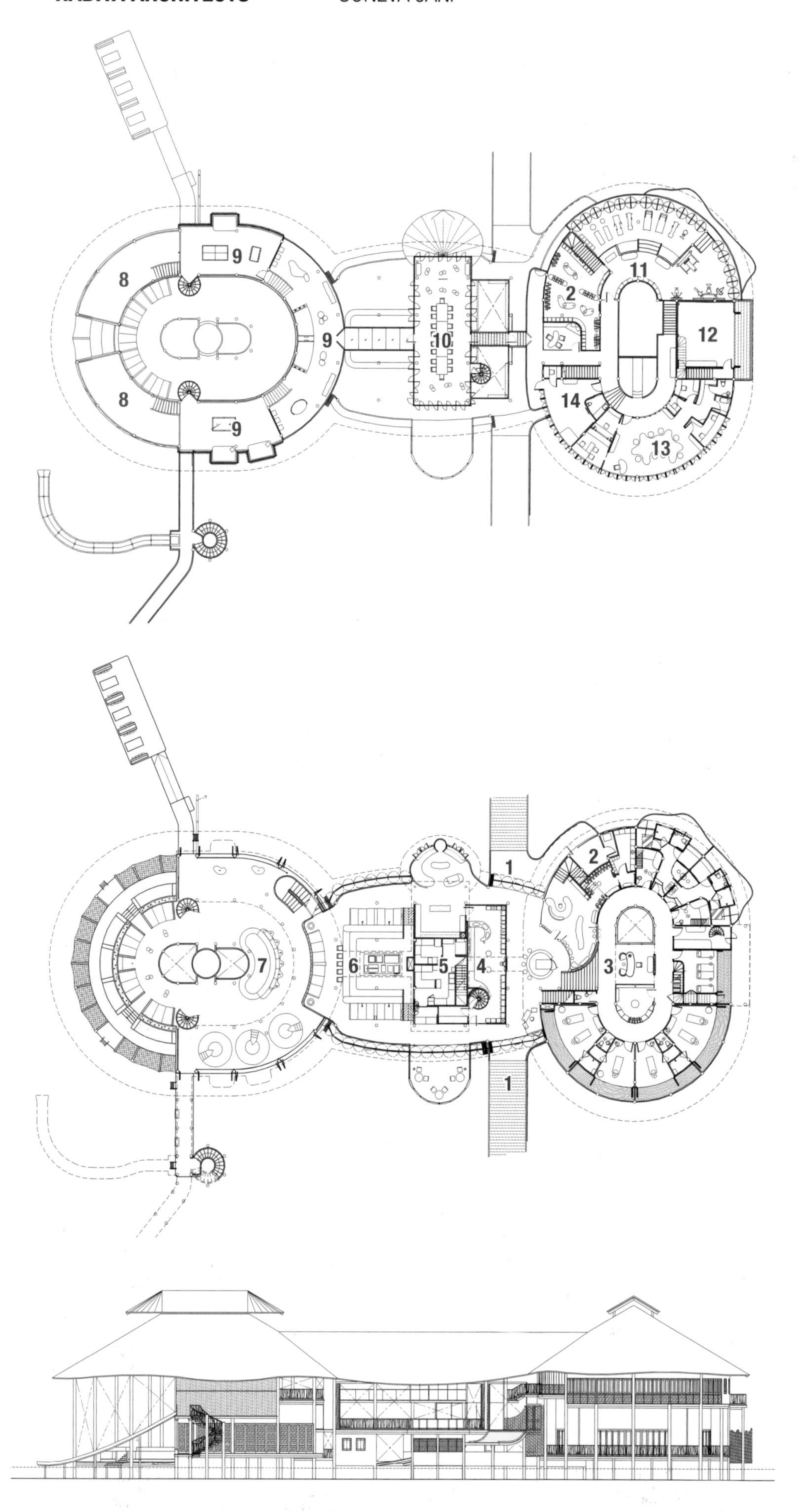

CLUB HOUSE

1 Entrance
2 Retail
3 Spa
4 Wine cellar
5 Kitchen
6 Open kitchen
7 Bar
8 Dining area
9 Library / game area
10 Wine-tasting room
11 Gym
12 Yoga
13 Office
14 Kids' den

0 5 10 20m

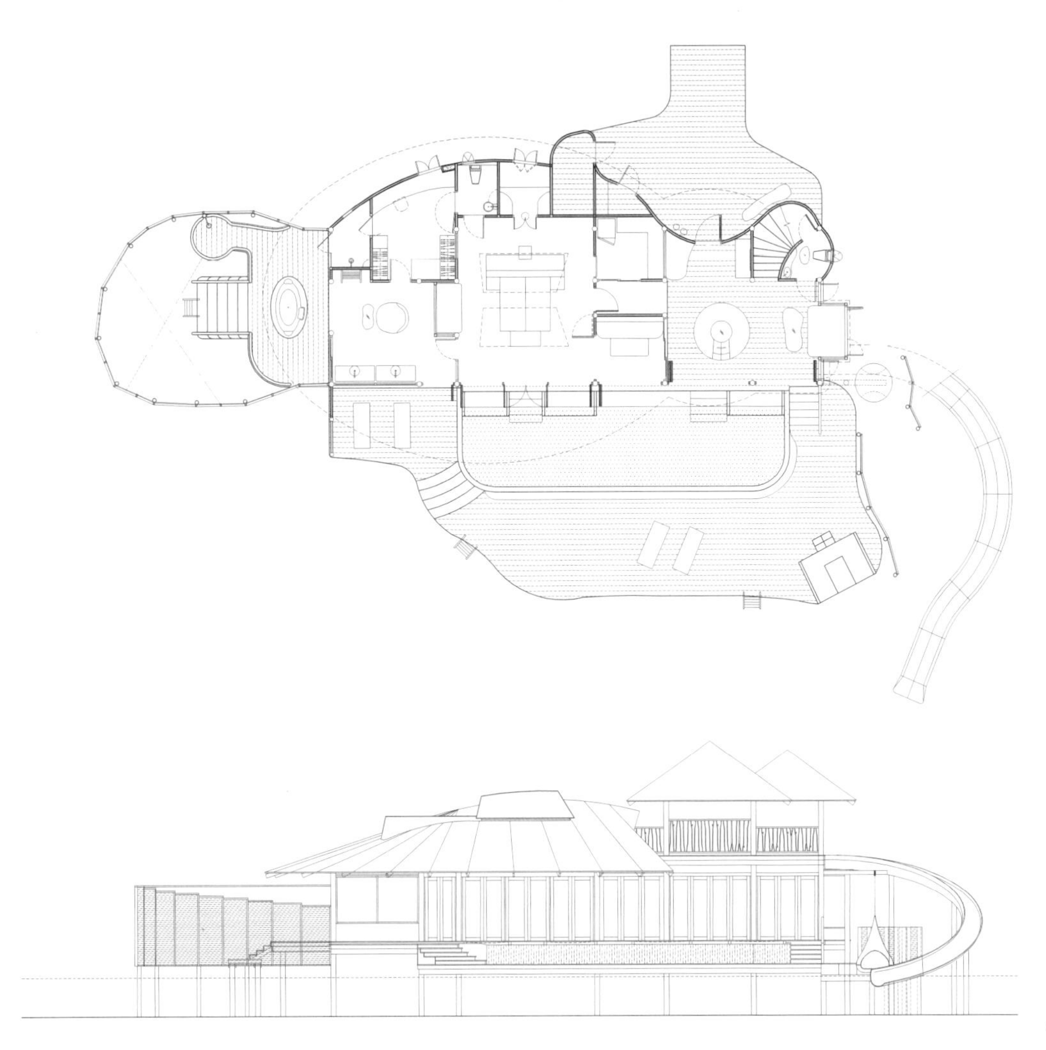

WATER VILLA (ONE BEDROOM)

0 5 10 20m

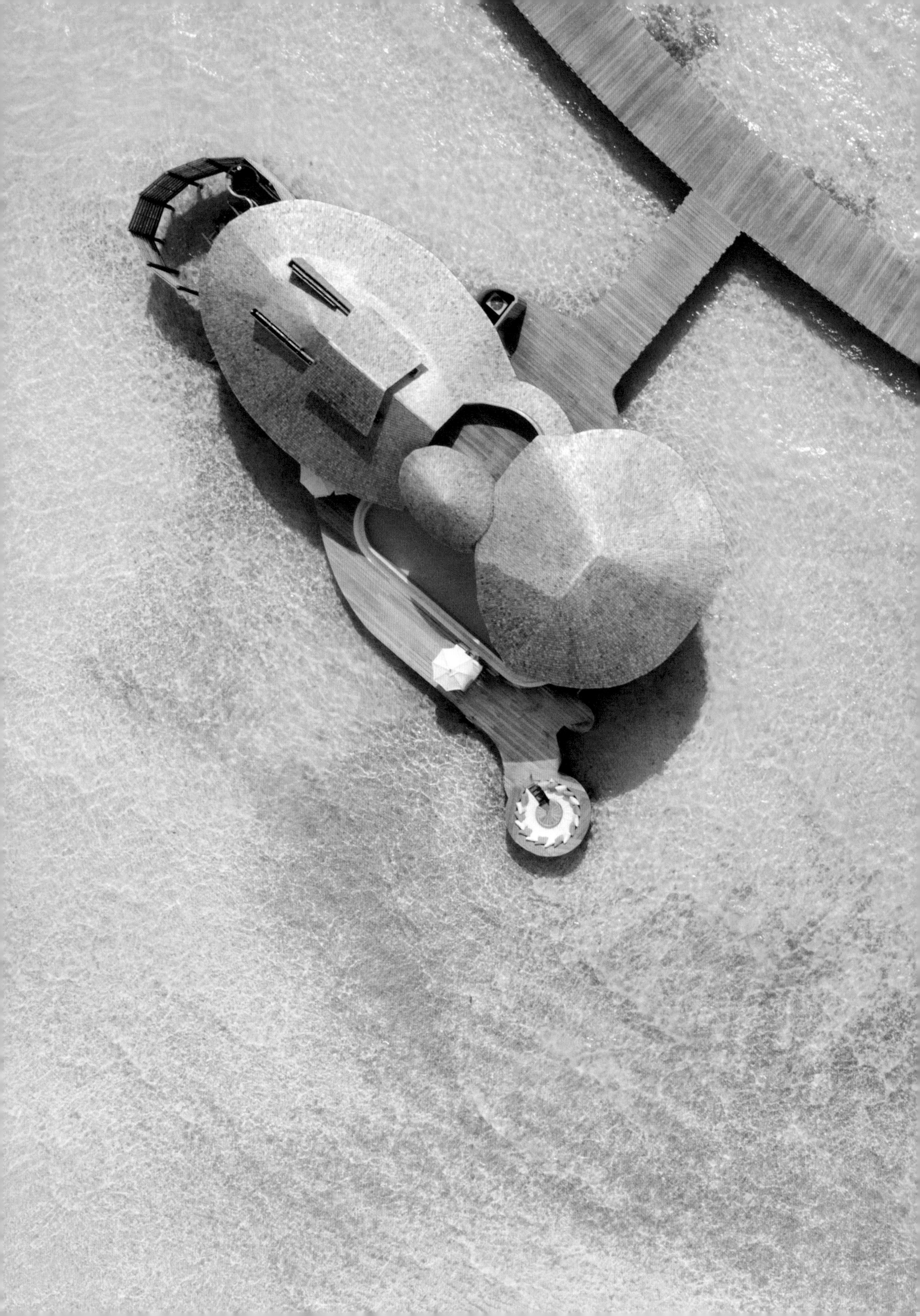

普吉岛迈考沙滩安纳塔拉度假会

ANANTARA MAI KHAO PHUKET VILLAS

Location Phuket, Thailand
Date 2008
Interior designer P49 Deesign & Associates
Landscape designer Bensley Design Studio

Anantara Mai Khao Phuket Villas are located next to the virgin sand beach and lush tropical vegetation of Sirinath National Park, near Phuket, Thailand.

The resort comprises 83 tropical pool villas and includes premium restaurants, bar, and leisure facilities. The pool villas are set in a village-like configuration around an expansive and richly landscaped tropical lagoon. The lobby, all-day restaurant, bar, and a destination spa form the heart of the resort, also focused on the lagoon.

The pool villas are composed in a cluster of four units, resembling a large traditional southern Thai house. Each unit has its own entrance and a private swimming pool for the guests' privacy and relaxation. This clustered arrangement provides a high-density solution on the 43 rai (17 acres) site, permitting views of the large lagoon and the resulting relaxed ambience of villas located alongside a canal. The floors raised high off the ground, the hip roof, and roof cement roofing tiles are all traditional elements of southern Thai architecture.

Guests enter the resort through the narrow driveway to the arrival courtyard, and the gable roof at the entrance opens out to the tranquil backdrop of the lagoon. From here, guests walk on a raised wooden pathway along the water garden to the reception and lobby pavilion. After checking in, guests can either walk or take a golf-cart to their villas. Access to the beach and beach facilities is via raised walkways. The beach restaurant and the infinity-edge swimming pool occupy the full beach frontage of the site and provide a panoramic view of the Andaman Sea. From here guests can enjoy the palm-lined beach with food and beverages on hand to restore energy levels after time spent outdoors.

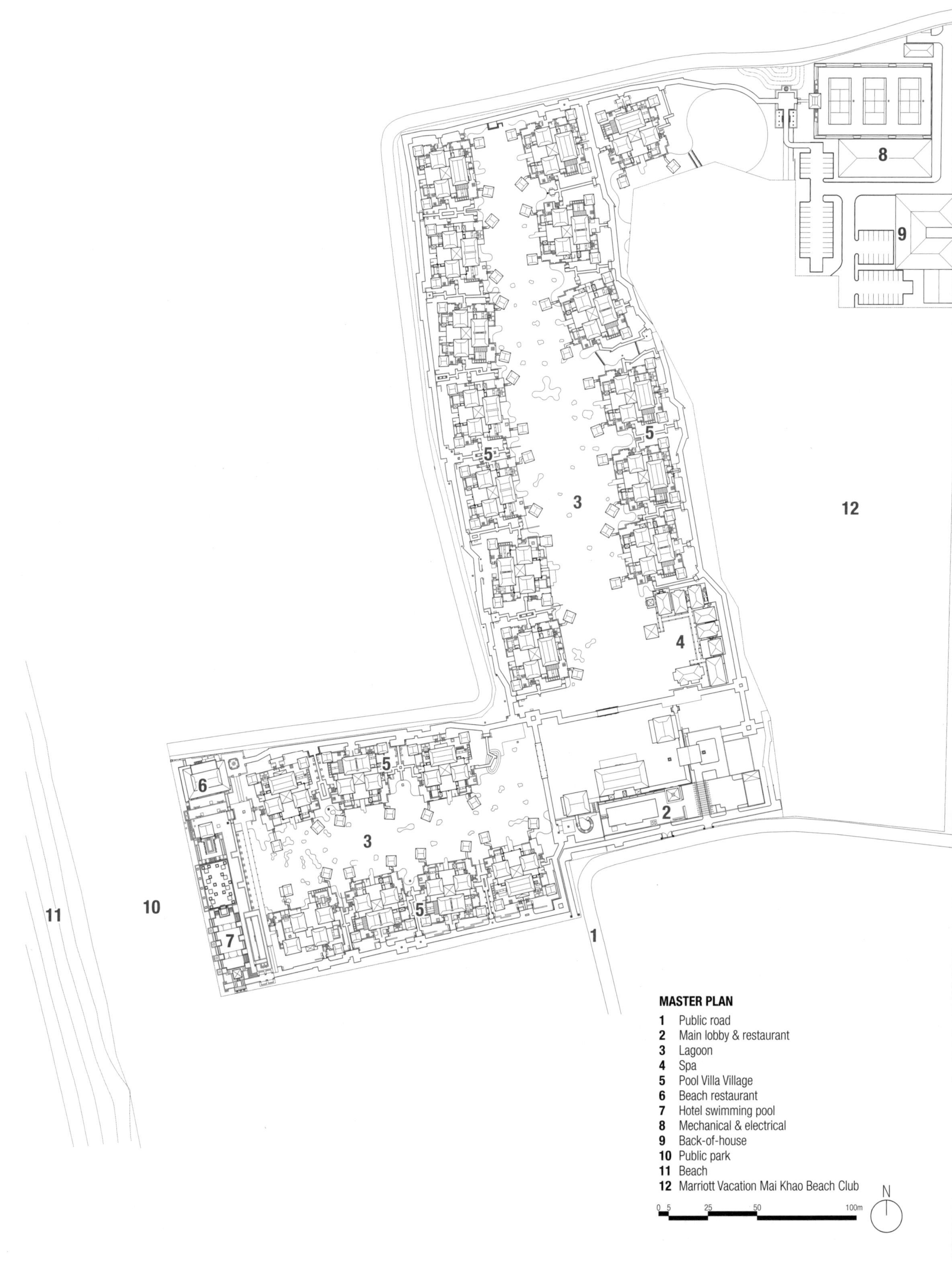

MASTER PLAN

1 Public road
2 Main lobby & restaurant
3 Lagoon
4 Spa
5 Pool Villa Village
6 Beach restaurant
7 Hotel swimming pool
8 Mechanical & electrical
9 Back-of-house
10 Public park
11 Beach
12 Marriott Vacation Mai Khao Beach Club

MAIN FACILITIES

1 Arrival court
2 Reception pavilion
3 Meeting facilities court
4 All-day dining
5 Display kitchen
6 Kitchen
7 Wine cellar
8 Lagoon

0 5 10 20m

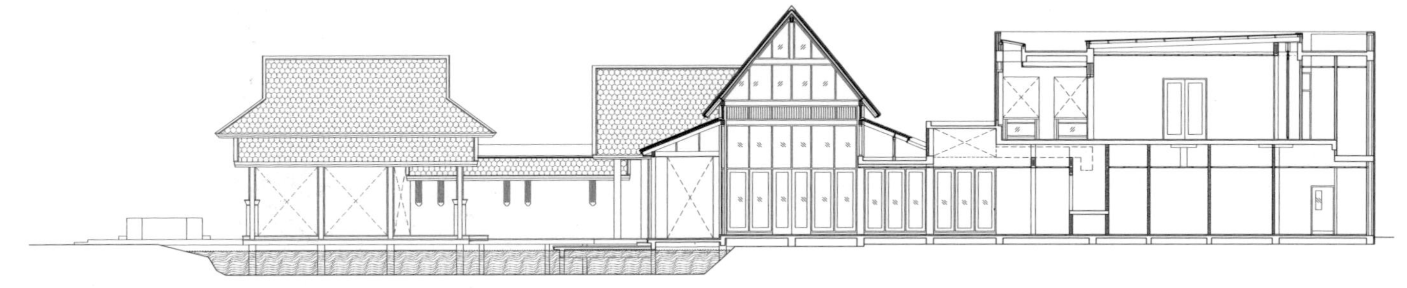

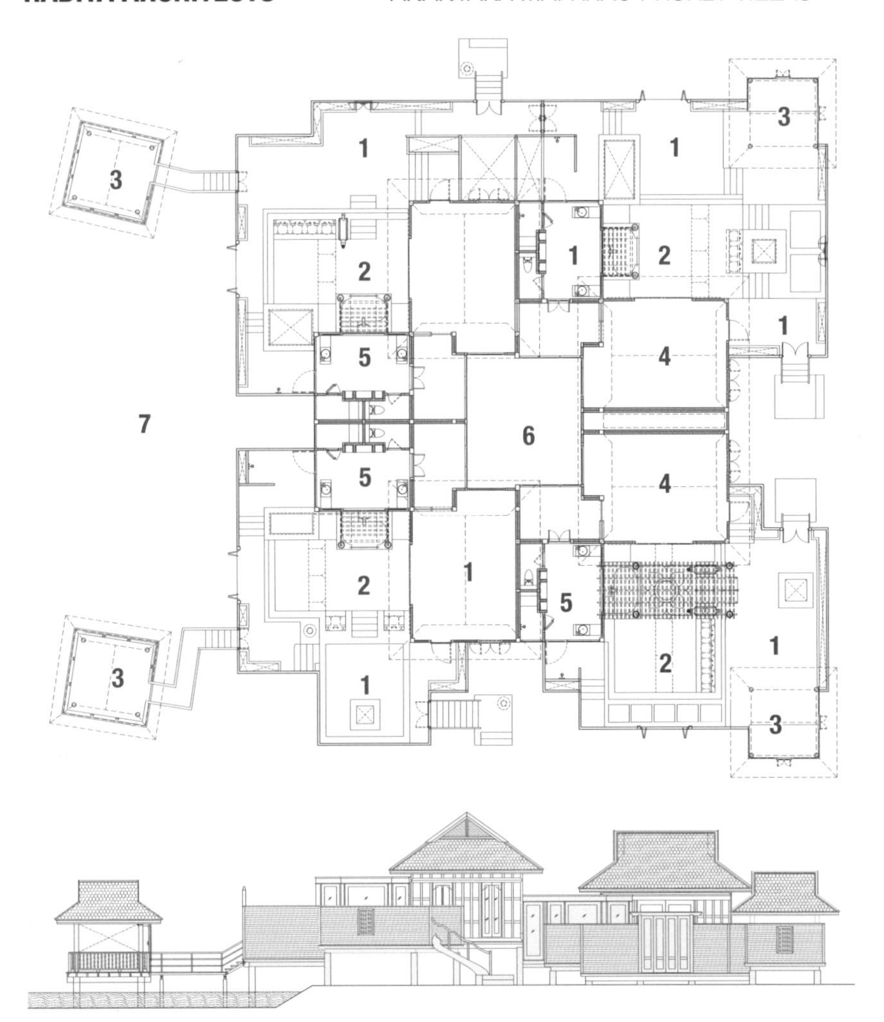

VILLA

1 Entry court
2 Swimming pool
3 Sala
4 Bedroom
5 Bathroom
6 Garden court
7 Lagoon

0 5 10 20m

帕岸岛拉沙南达安纳塔拉别墅度假村

ANANTARA RASA NANDA KOH PHANGAN VILLAS

Location Koh Pha Ngan, Surathani,Thailand
Date 2009
Interior designer P49 Deesign & Associates
Landscape designer Inside Out Design

Koh Pha Ngan is a sister island of larger Koh Samui, located in the Gulf of Thailand. While it is famous for its full moon party, the island retains much of its local character and natural, unspoilt beauty. Thong Nai Pan beach, with its white sand and clear blue water, has always been a popular beach for tourists. In fact, the site of this luxury resort was previously occupied by backpacker accommodations.

Generally, arrival to the resort is by speedboat to the beach. However, the main reception has been located to provide the same level of guest arrival experience as those arriving by ferry and then by road across the island. Despite it being a world-famous tourist destination, Koh Pha Ngan retains a strong rural feel. The concept of this development was to reflect this character throughout the resort environment. Hence Anantara Rasa Nanda has a very organic, laid-back feel.

Arriving from the beach, only some of the resorts roofs and beach seating are visible. Otherwise, all that the visitor sees is its nearly 800-foot-long (240-meter-long) beach frontage, coconut palms, and rainforested hillside behind. Guests then enter the site adjacent to the beachfront restaurant and infinity pool, stepping up gradually to the reception area.

Sand from the beach continues right into the restaurant's double-height external dining area. Right alongside is the pool bar with thatched roof. Here guests immediately experience the casual feel of the village, before proceeding to the cool reception area and lounge set back further into the trees.

Through the road entrance gate and across this local-use road, visitors enter the resort's spa. Here the site transitions from flat beachside to tropical hillside. The spa has an entirely different feel to the villas and main facilities. It is dark and subdued, with treatment rooms set among rockpools and dense vegetation.

Back on the beachside, the site enjoys a relatively unusual parallel orientation to the beach. Of the overall development of 26 villas, a total of 17 occupy premium beachfront locations. The layout of the central access path parallel to the beach allows proximity of all villas to the sights and sounds of the sea. Continuing the overall rural village theme, the four different villa types provide visual variety in terms of scale and backdrop. The villas are both single- and double-story, and all incorporate traditional southern Thai architecture consisting of a concrete tile roof, thatch, and the selective use of natural timber.

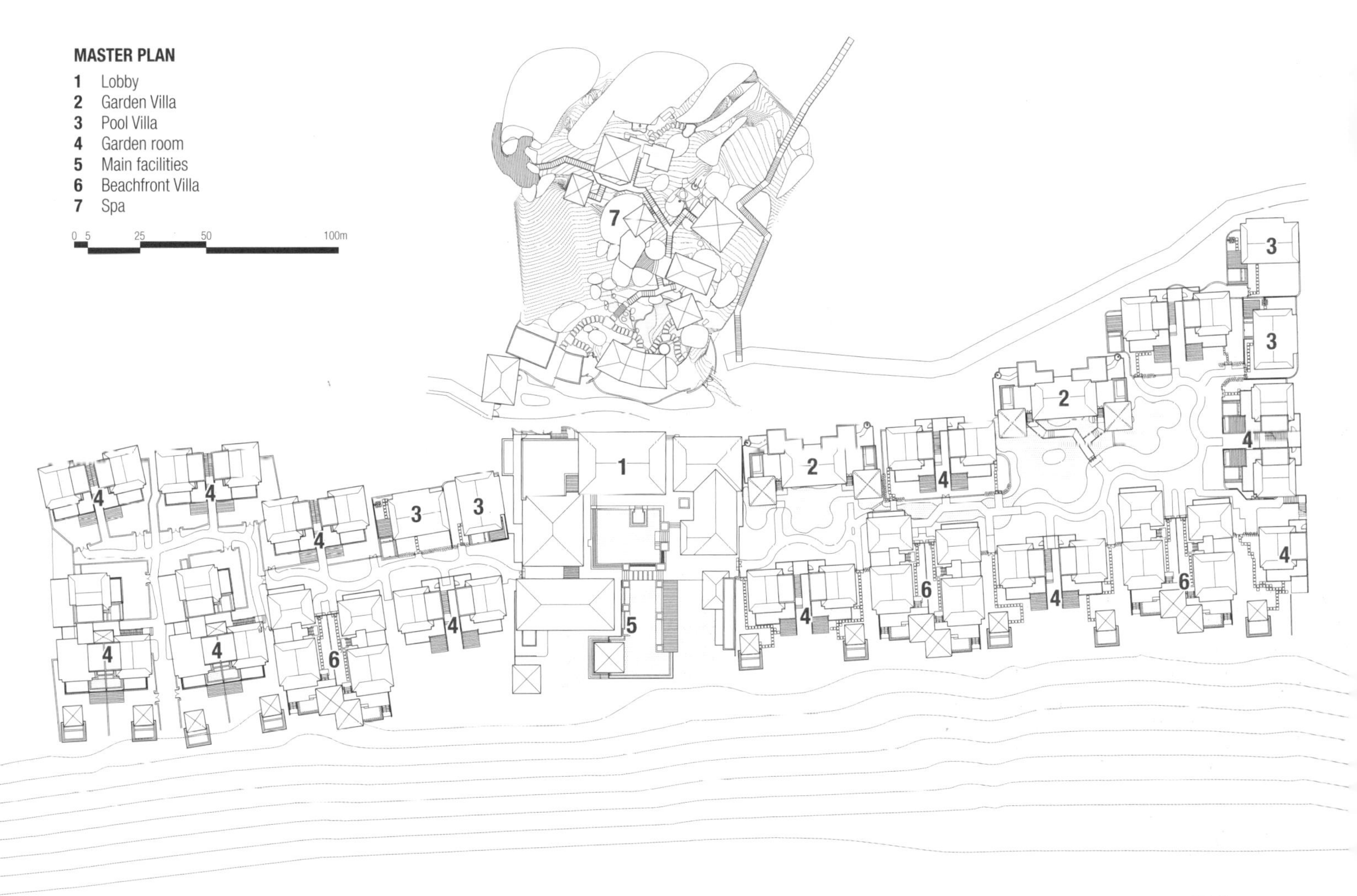
MASTER PLAN
1 Lobby
2 Garden Villa
3 Pool Villa
4 Garden room
5 Main facilities
6 Beachfront Villa
7 Spa
0 5 25 50 100m
1
2
3
4
5
6
7

MAIN FACILITIES
1 Reception
2 Front office
3 Lounge
4 Kitchen
5 Buffet
6 Dining
7 Pool
8 Bar
9 Ocean sports
10 Shop
0
5
10
20m

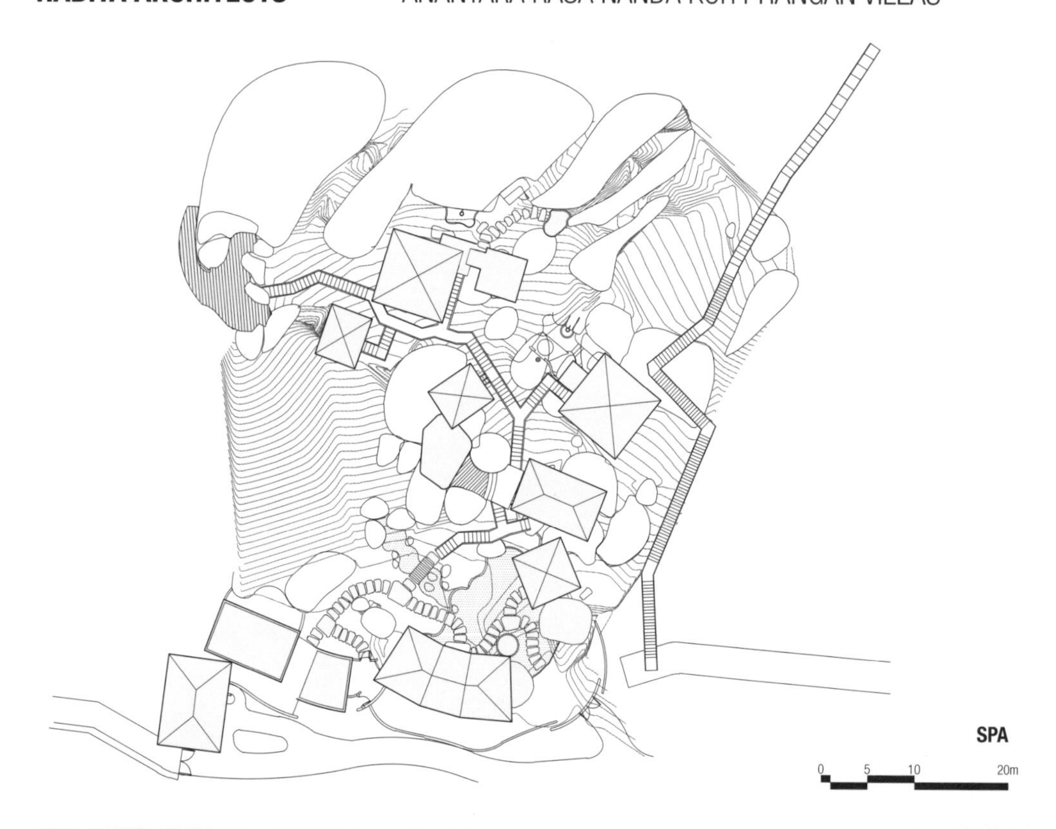
SPA
0
5
10
20m

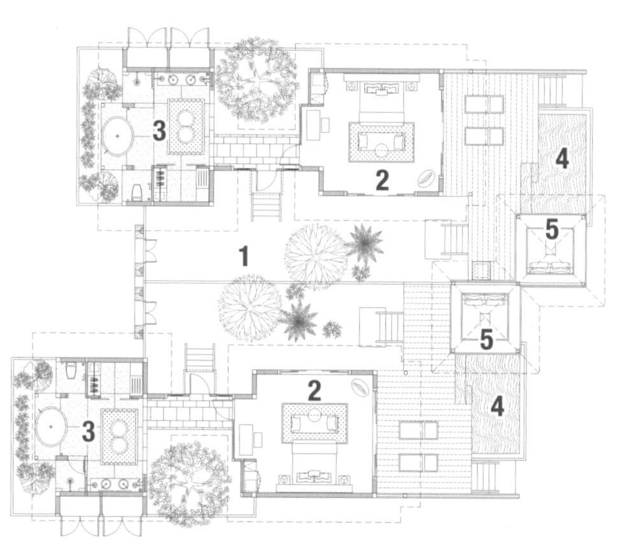

BEACHFRONT VILLA

1 Courtyard
2 Bedroom
3 Bathroom
4 Pool
5 Sala

普吉岛拉扬安纳塔拉度假村

LAYAN RESIDENCES BY ANANTARA

Location Phuket, Thailand
Date 2016
Interior designer Jaya International.
Landscape designer Intaran Design

The Layan Residences by Anantara is an exclusive development of 15 private villas set into the hillside directly above the Anantara Layan resort on Phuket's western coast. Accommodating the client's programmatic requirements (ranging from two-bedroom up to very ample nine-bedroom villas, all with pools and spacious living spaces), without effacing this otherwise untouched rainforest backdrop to the resort below, represented a considerable challenge. Furthermore, we had to achieve this high standard of residential luxury on a site doubly constrained by steep topography and zoning laws, restricting the maximum building height to 20 feet (6 meters).

The building height constraint made it obligatory to use flat roofs throughout the development, and to locate living spaces at various levels. Meanwhile, multiple levels specifically undermined the expansive experience of space envisaged for such grand retreats.

We approached this challenge by dividing the villas into repeated components—master bedroom, living and dining pavilions, and a fitness space with pool integrated with the roof. As well as benefiting from a panoramic view of the bay, which residents using the gym can enjoy, this space is illuminated by skylights set into the floor of the swimming pool overhead. The slope of the site was such that the roofs of these various components serve as outdoor living space close to the floor level of the component positioned behind, and on the hillside. Each villa was then individually designed, with the respective components arranged according to the contours on which the villa was situated. A palette of natural materials—stone and timber—was used throughout. A three-tiered roof motif, with indented corners, introduces a Thai element to the modern flat roof design.

Principal living spaces such as the living room, dining room, and bedroom components, share the same floor level along existing contour lines, separated by deck spaces and covered walkways that augment the connection of these spaces. Extending toward the sea from these spaces in each villa is an extensive outdoor living plinth, accommodating pool, sala, ample sun lounger space, and landscaping. The slope of the site is such that the neighboring villa in front is set well below the vista of the bay. Green roofs further minimize visual intrusion.

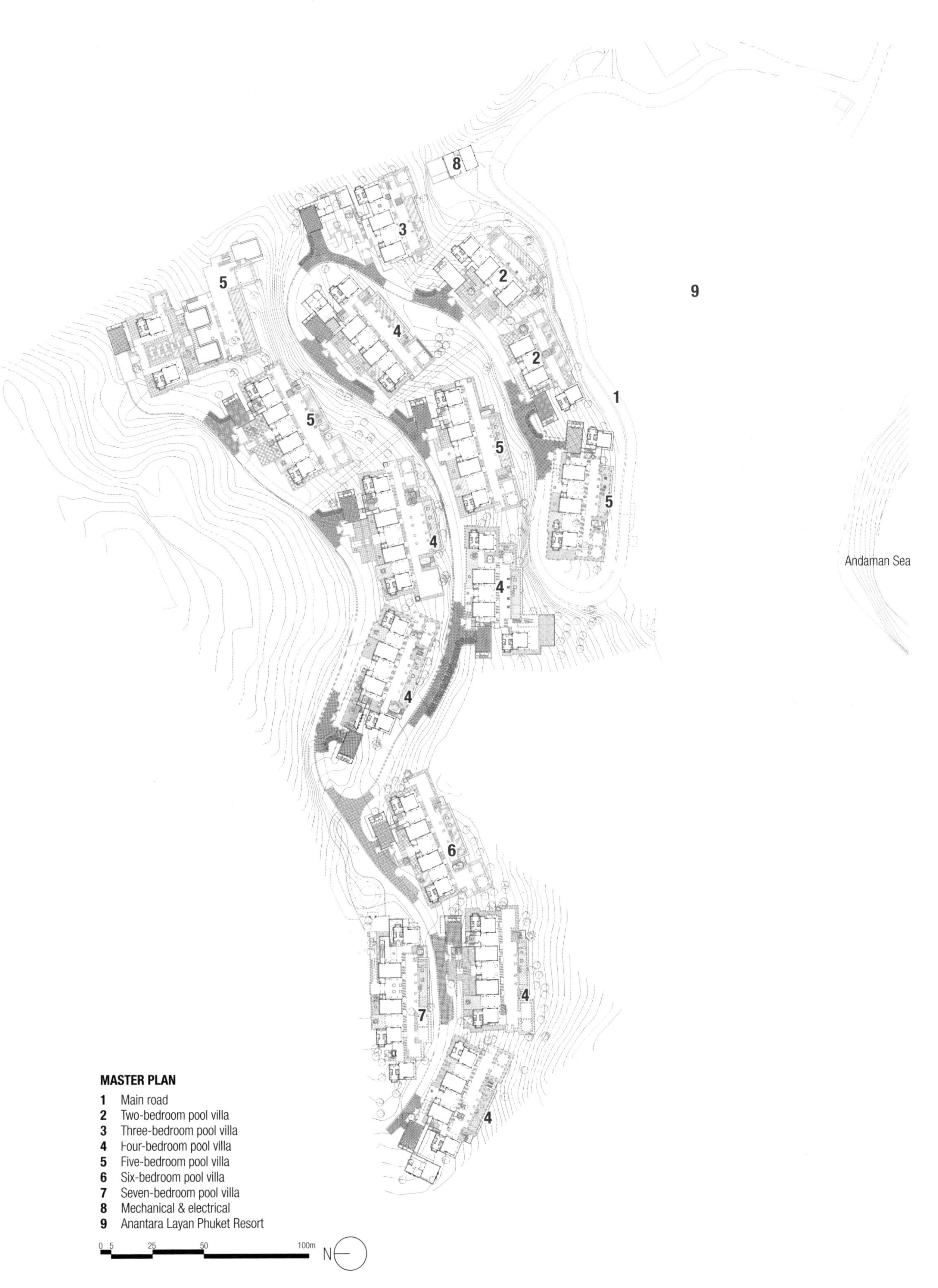
8
3
5
2
9
4
2
1
5
5
5
4
4
Andaman Sea
4
6
4
7
4
MASTER PLAN
1 Main road
2 Two-bedroom pool villa
3 Three-bedroom pool villa
4 Four-bedroom pool villa
5 Five-bedroom pool villa
6 Six-bedroom pool villa
7 Seven-bedroom pool villa
8 Mechanical & electrical
9 Anantara Layan Phuket Resort
0 5 25 50 100m
N

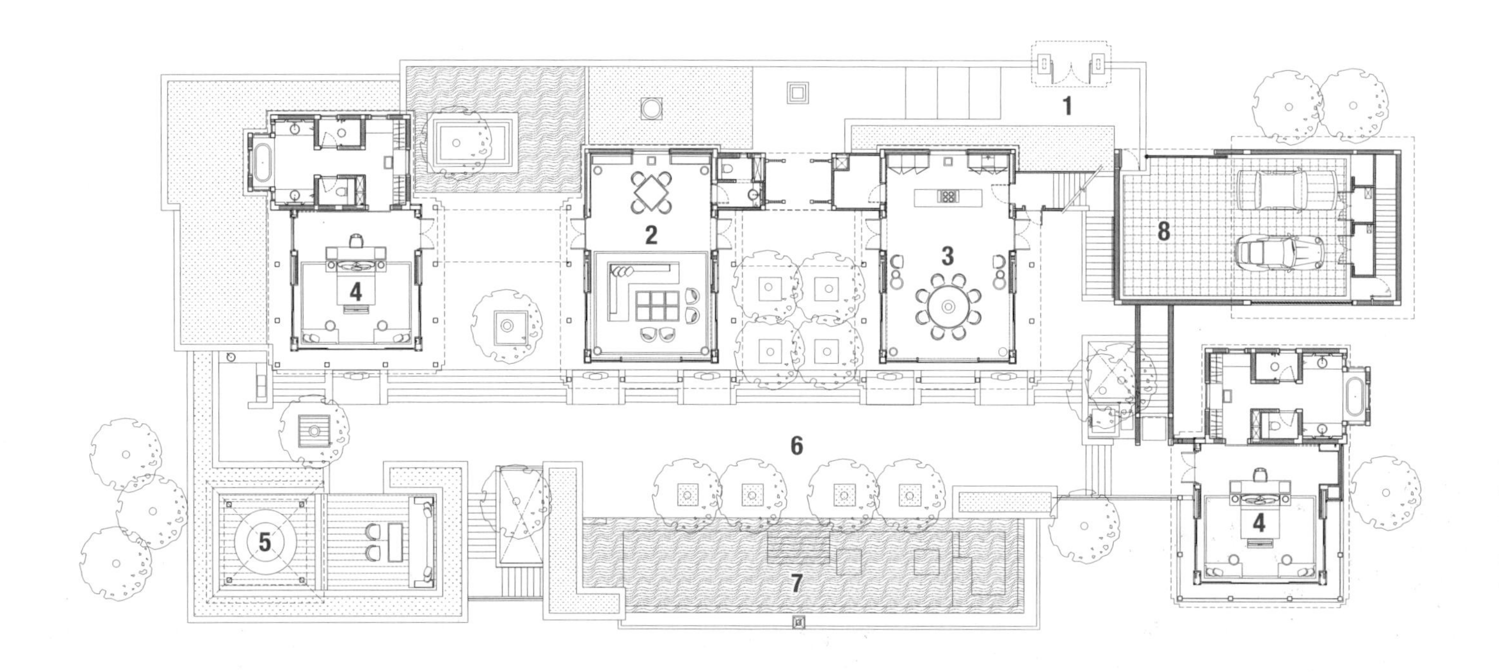

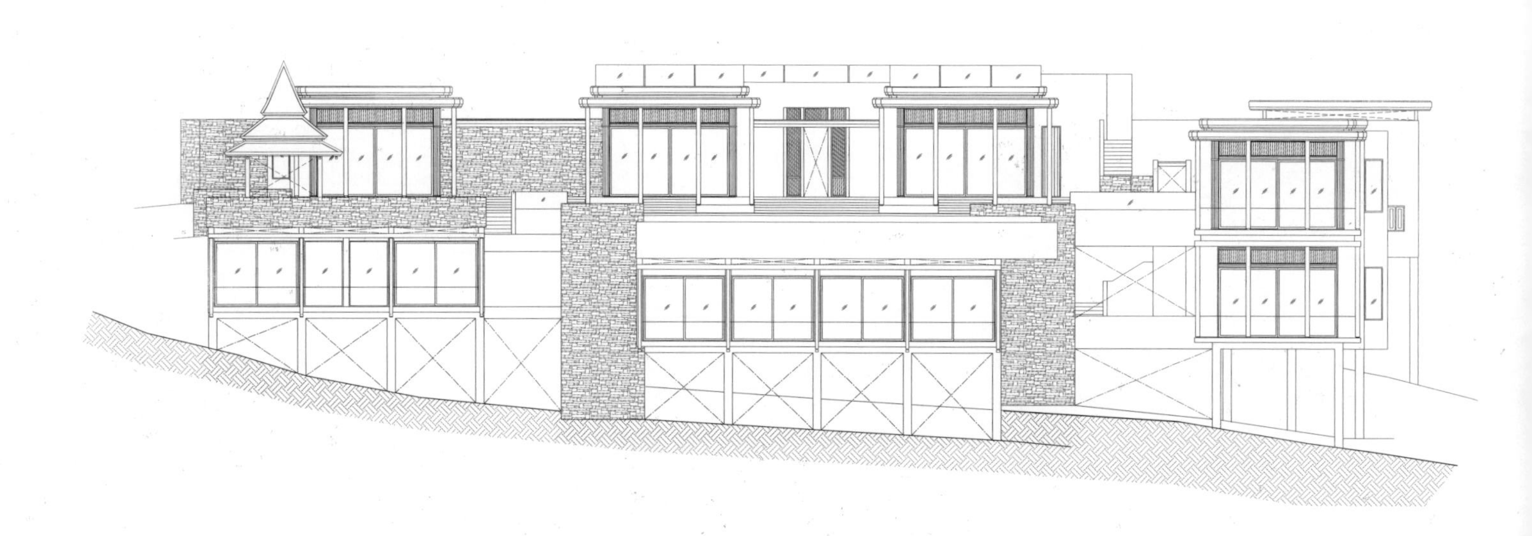

RESIDENTIAL VILLA

1 Entry
2 Living room
3 Dining room
4 Master bedroom
5 Sala
6 Pool deck
7 Swimming pool
8 Parking

0 5 10 20m

普吉岛芭东海滩智选假日酒店

HOLIDAY INN EXPRESS PATONG

Location Patong Beach, Phuket, Thailand
Date 2012
Interior designer P49 Deesign & Associates
Landscape designer Inside Out Design

This 280-room hotel is located near Patong Beach in Phuket. It is one of the first Holiday Inn Express (HIEX) developments on the island, with a resort rather than city hotel atmosphere.

The existing land was a coconut farm. The hotel is inserted between the coconut palms to generate a relationship between human space and nature, with guestrooms facing a farm in which coconut palms are the main component. The hotel is conceived as sitting within a grove of coconut palms.

As HIEX is a budget hotel, the project required a simple design that was functional and whose architectural elements met cost constraints. The concepts behind these architectural elements are as follows:

Purple color: this color not only happened to be a personal favorite of the client, but is appropriate to the area, representing the vibrant nightlife of Patong Beach. The exact color that was selected has become emblematic of the hotel.

Wall texture: a unique pattern was created and embossed onto concrete surfaces throughout the project, used in combination with areas of plain surface. This composition also became important to the project's overall identity.

Timber cladding: timber was used introduced to interior areas intimate to guests. The building exterior also features areas of timber cladding and components, connecting to the nature of the coconut grove. This is seen in the columns at ground-floor level, as well as the soffit and party wall to guestroom balconies. The warm timber color contrasts with the hardness of the gray concrete wall and complements the purple hue, bringing a relaxed atmosphere to the development.

These design elements were used in various combinations throughout the hotel. Together they celebrate both the joyful exuberance of Patong and the concept of the tranquil hotel within the coconut grove.

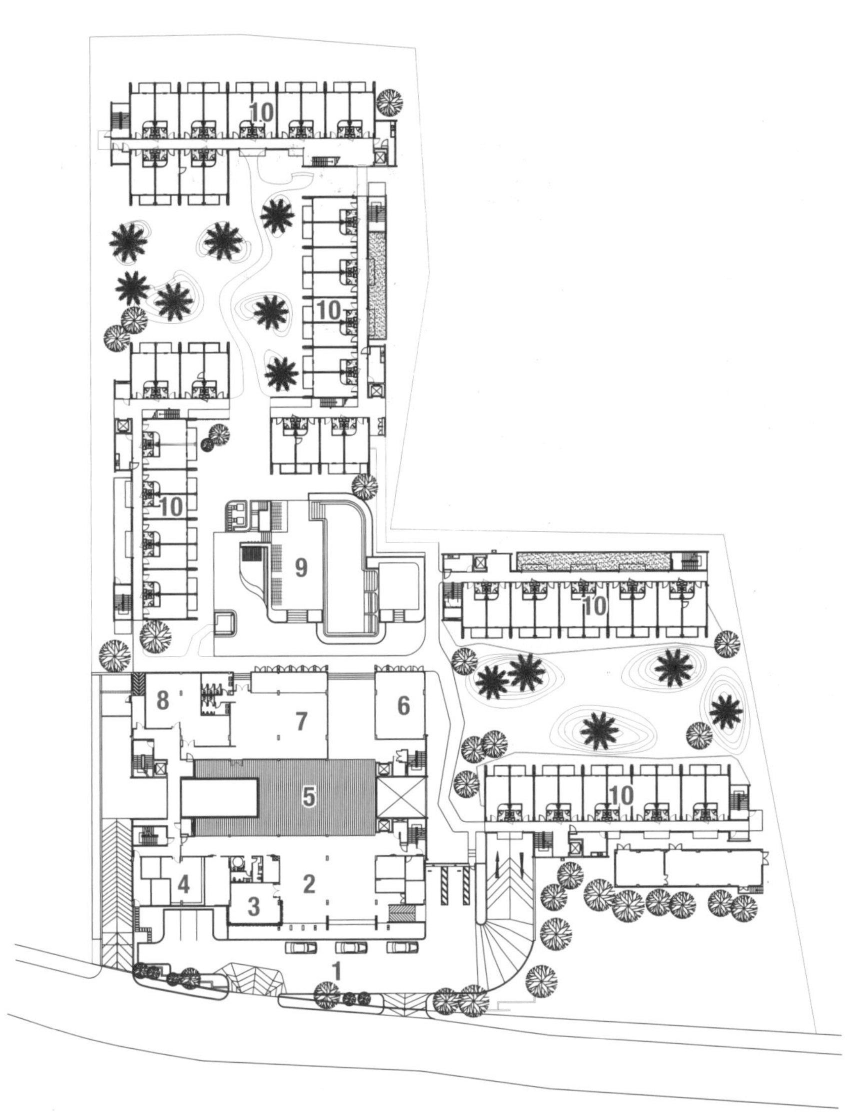

FLOOR PLAN

1 Drop-off
2 Lobby
3 Meeting room
4 Office
5 Reception
6 Library
7 Restaurant
8 Gym
9 Swimming pool
10 Guestroom

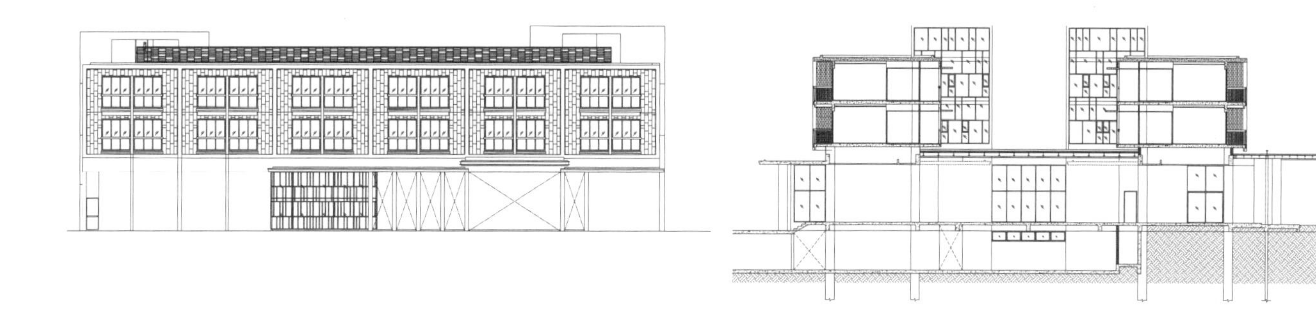

FAÇADE

0 5 10 20m

EXIT

普吉岛阿卡迪亚奈通海滩铂尔曼度假酒店

PULLMAN PHUKET ARCADIA

The Pullman Phuket Arcadia is located overlooking the Andaman Sea on the west coast of the island. It has an unusually spectacular setting, with high cliffs, large tropical trees, and three natural waterways flowing down the hillside. By creating several descending reservoirs, the drama of the site was emphasized and made more habitable for this sizable development.

Designed in tropical contemporary style, the 227-room hotel includes seven pool villas, as well as main facilities buildings. All buildings overlook each other, located to optimize sea views, with most buildings having flat roofs to minimize obstruction.

The sense of arrival is an outstanding feature of this hotel. A tunnel through rock brings guests to the drop-off, who then walk across a bridge with cascading ponds underneath. The bridge leads to steps that lead up to the lobby. Here the magnificent panoramic view reveals itself. As the lobby is on the second floor, the reflecting infinity-edged pool dispenses with the need for a safety rail and allows an obstructed view over the Andaman Sea.

The lobby and restaurant are located dramatically high above the sea. The free-form infinity edge of this cluster mimics the waves of the bay below. Here three areas of lounge seating areas are sunken into the reflecting pond, providing a memorable setting from which to experience the view.

In the main facilities buildings, strong gable and hip roof forms reflect local character. The timber roof structure is revealed in the canopy over the covered bridge. This continues right up into the lobby, as though the entire space is a single open 'sala'—a common space in traditional Thai houses.

Corridors and stairs are located away from guestrooms, connected by bridge to a private vestibule for each room with an entrance gate and main door. This separation also creates a natural light well to the guests' bathroom.

The landscape design pays close respect to the existing site conditions. Large rubber and cashew trees are retained to provide a flourishing verdant jungle atmosphere. These combine with newly formed waterfalls and lagoons, interacting closely with the buildings throughout the site.

Location Naithon Beach, Phuket
Date 2012
Interior designer P49 Deesign & Associates
Landscape designer P Landscape

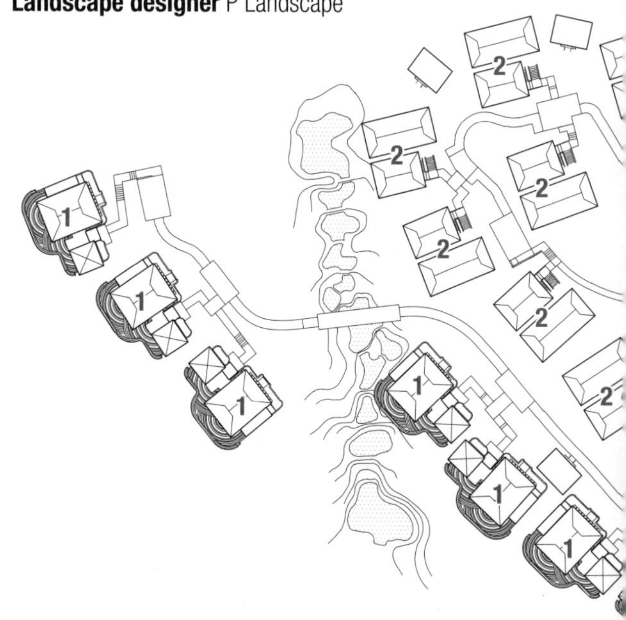

MASTER PLAN
1 Pool villa
2 Cluster villa
3 Hotel room
4 Lobby
5 Main pool
6 Parking
0 10 50 100m
N

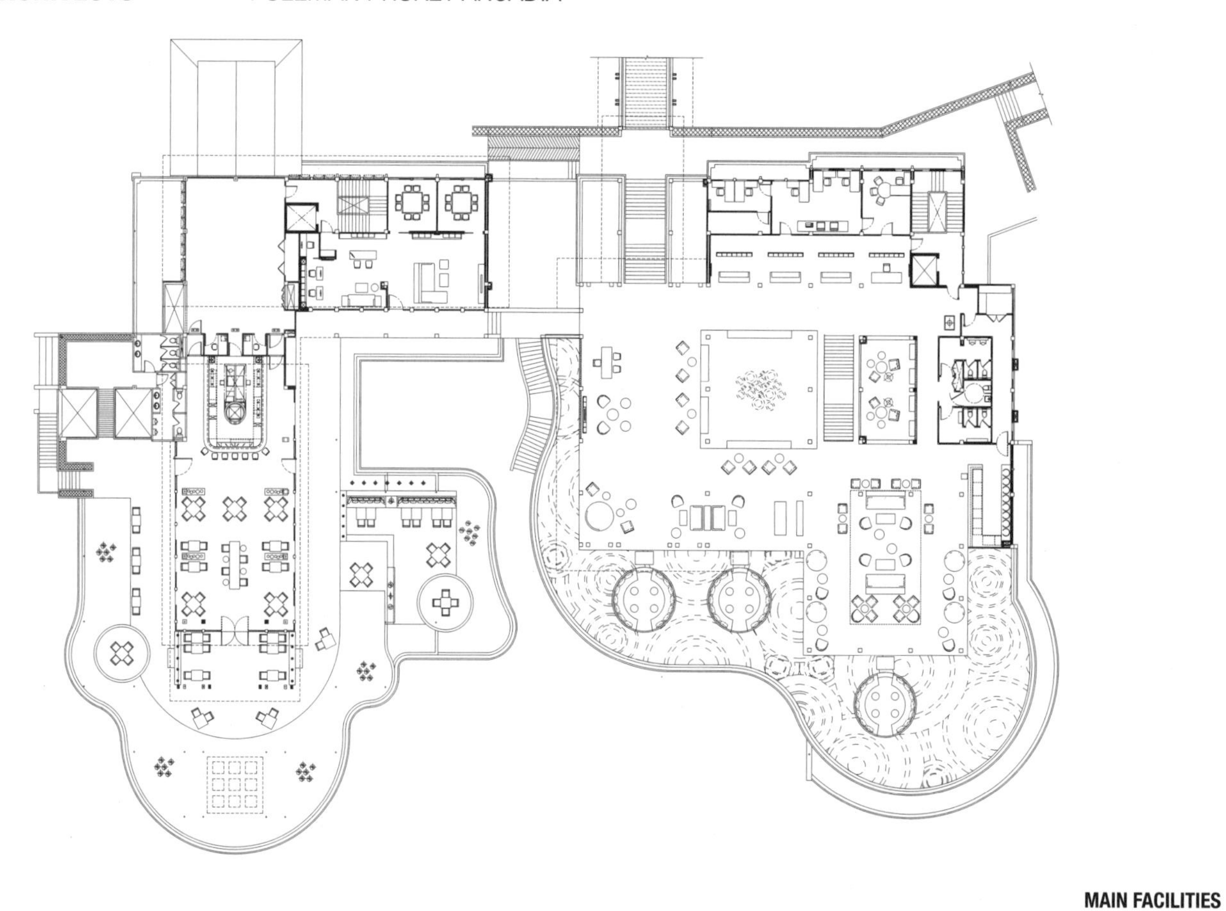

MAIN FACILITIES

0 5 10 20m

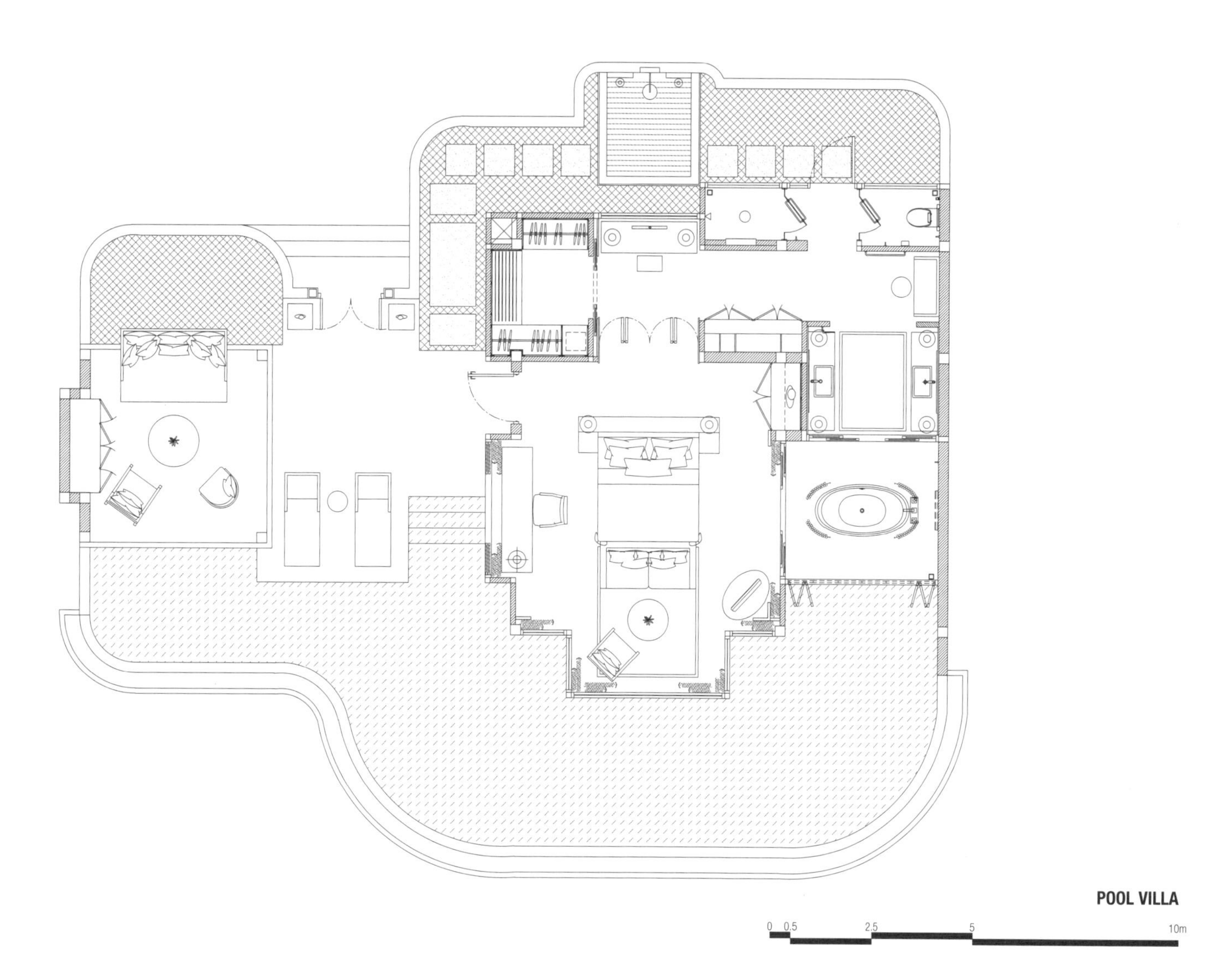

POOL VILLA

0 0.5 2.5 5 10m

曼谷麦卡桑美居酒店

MERCURE BANGKOK MAKKASAN

Location Bangkok, Thailand
Date 2017
Interior designer Innia
Landscape designer Inside Out Design

Mercure Bangkok Makkasan is a 4-star hotel in the center of Bangkok. It is located next to Makkasan Airport Link station, which is an important transit hub to Suvarnabhumi International Airport. The 25-story hotel has 184 guestrooms, as well as an all-day-dining restaurant, wine bar, swimming pool, gym, and conference facilities.

The core strength of the hotel is its ready access to many significant visitor amenities, local trains, BTS (skytrain), highway, and Bangkok's CBD. Given its location at this infrastructural hub, it was important to give the tower a strong external identity.

We set out to create a unique façade pattern by twisting guestroom windows at different angles. This gives the hotel a distinctive rhythmic visual identity, while also lending the hotel a different character when seen from different angles.

The twisted windows were initially conceived to provide each room with a different perspective of Bangkok. The purple color of the exterior façade expresses the international Mercure brand.

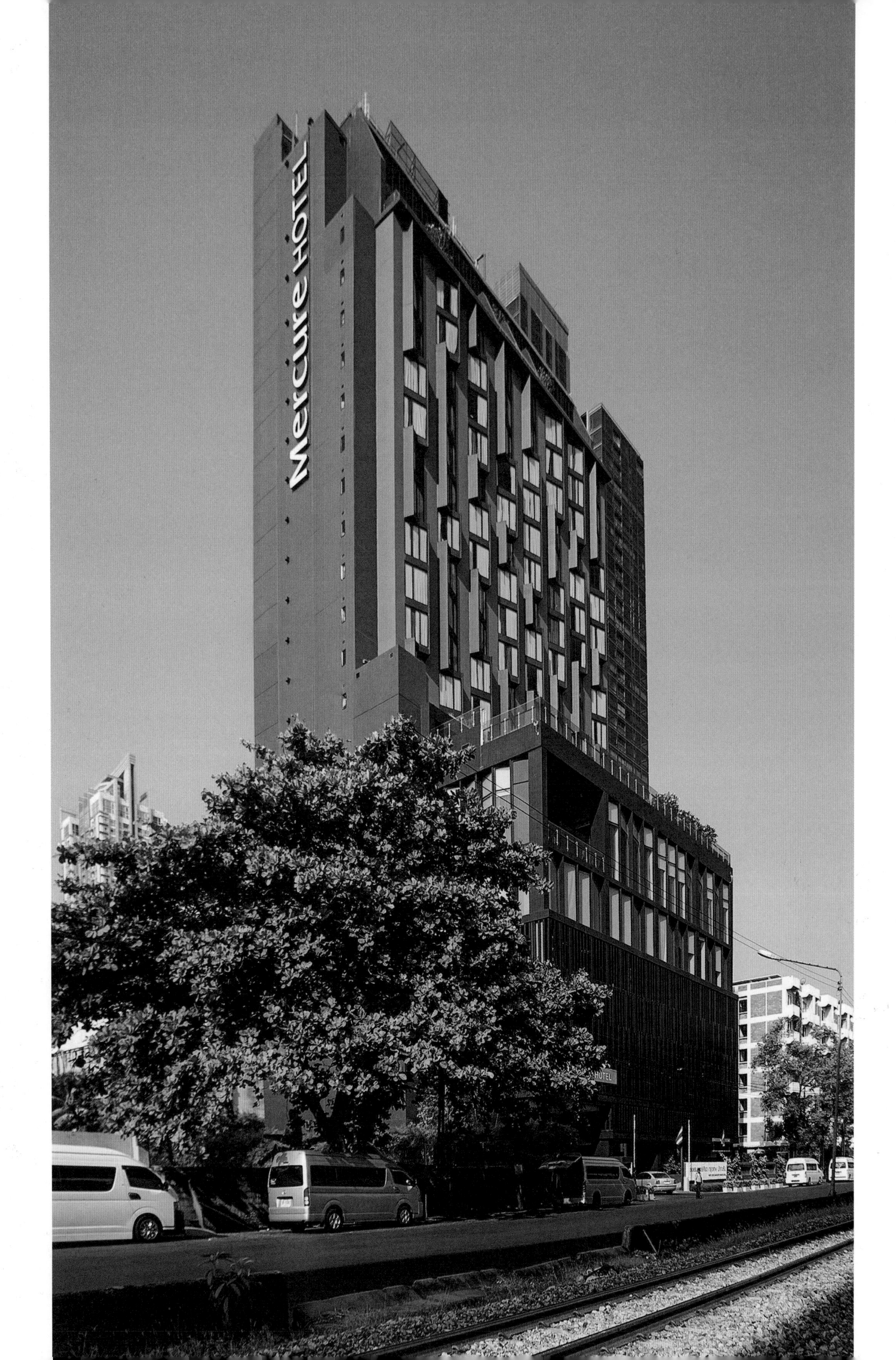
Mercure HOTEL

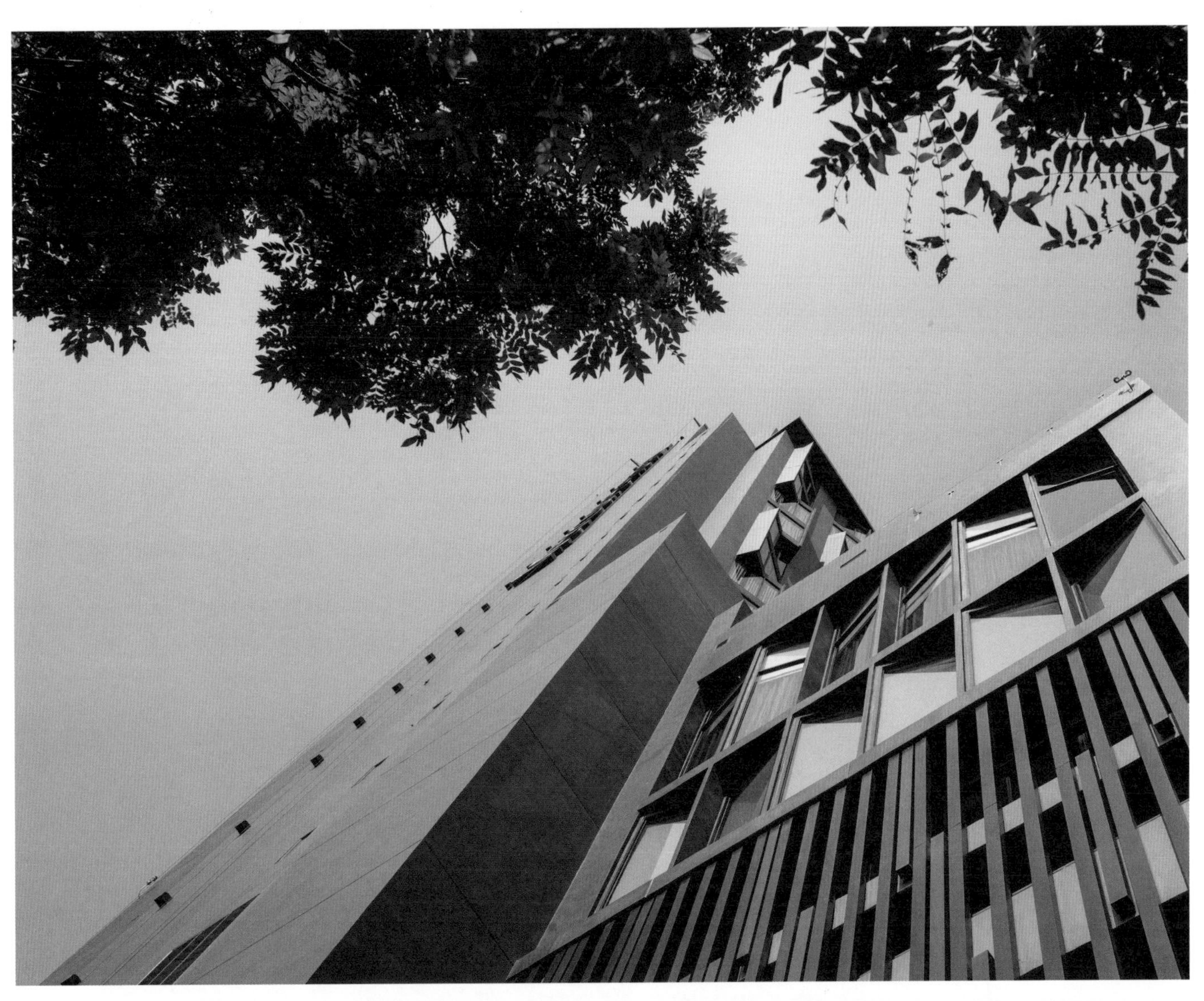

Mercure HOTEL
PARKING
โรงแรมเมอร์เคียว กรุงเทพ มักกะสัน
MERCURE BANGKOK MAKKASAN

Mercure HOTEL

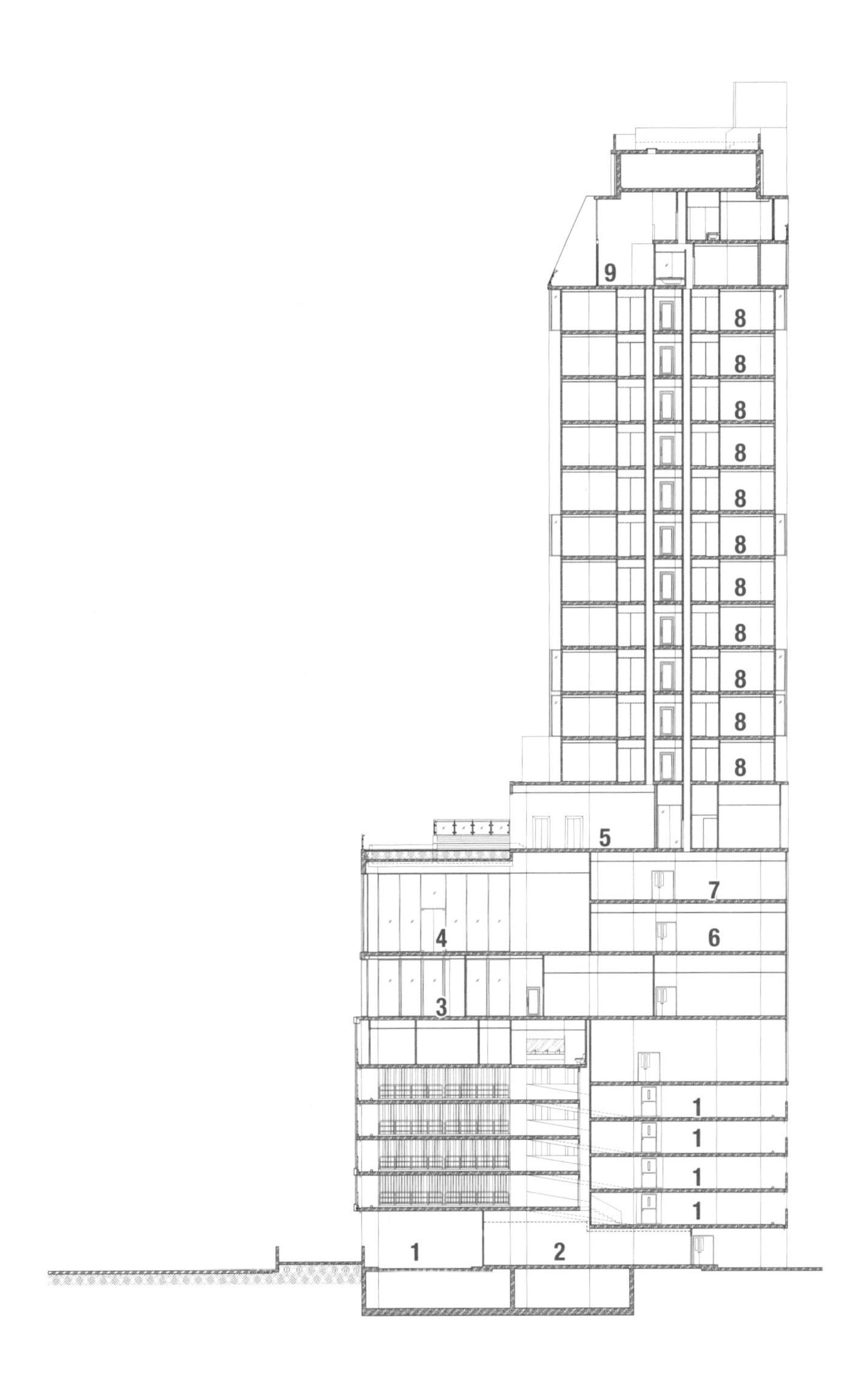

SECTION

1 Parking
2 Back-of-house
3 Meeting room
4 Dining room
5 Pool bar
6 Kitchen
7 Office
8 Hotel room
9 Penthouse room

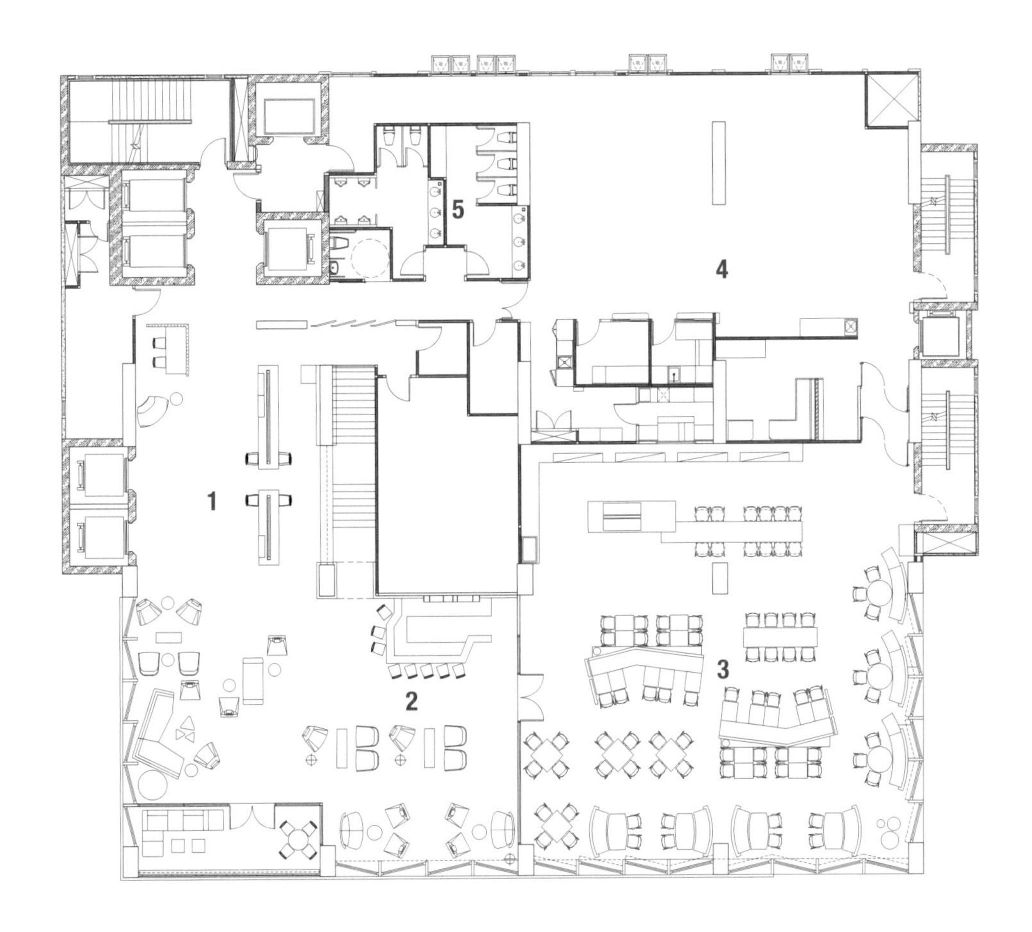

LOBBY FLOOR PLAN

1 Reception
2 Lobby
3 Dining room
4 Kitchen
5 WC

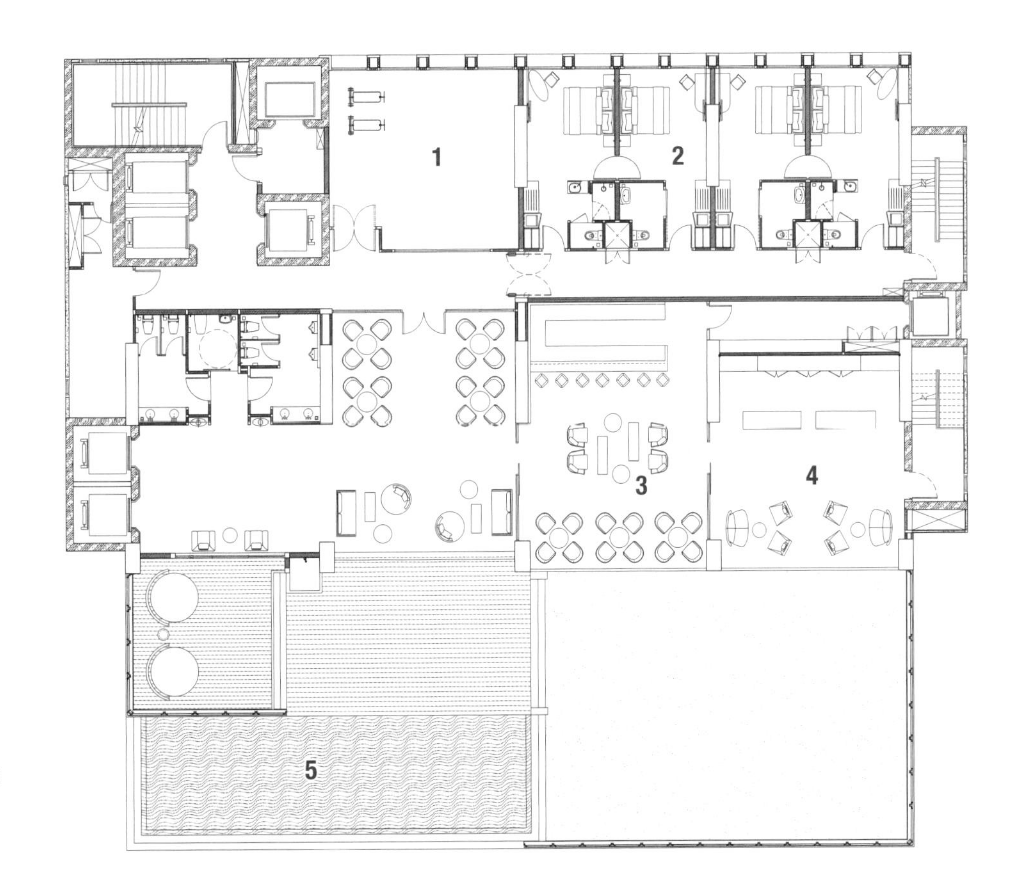

FACILITIES FLOOR PLAN

1 Fitness
2 Hotel room
3 Pool bar
4 Wine cellar
5 Swimming pool

0 1 5 10 20m

甲米碧玛莱度假村
PIMALAI RESORT AND SPA

The Pimalai Resort and Spa was developed in two separate phases, divided by a public road. The seaward site of the site was established first. Four years later, by which time the resort had acquired a very good reputation, work on the landward and more hilly side of the site commenced. Located on the western coast of Koh Lanta, the resort enjoys magnificent views over the Andaman Sea, as well as direct frontage onto a fine white-sanded beach.

During the regular season, guests arrive by boat directly onto the beach in front of the resort, keeping true to its island location. During the monsoon season, guests arrive at the island's main pier and transfer to the resort by land.

The arrival area and reception lobby are located to provide a sweeping panoramic view of the sea and hence provide guests with a breathtaking first impression.

In the first phase, villas were clustered into groups of four to sit gently into the terrain and avoid removal of any existing trees on the site. The style of these two-story villas is a homely tropical southern Thai style, with hipped clay tile roofs. A beachfront pool and restaurant was also included in this initial phase.

The second phase was located up on the hill in among many large, tropical trees. To compensate for the relative distance from the beach, each villa has its own pool. Also, building on the reputation and popularity of the first phase, these villas are all stand-alone units and are generally designed to be more luxurious than the first-phase accommodations. Also included in the second phase were additional main facilities, including a restaurant and lounge.

The architecture of the second phase also featured more complex roof forms and finer detailing. Reflecting not only the more luxurious nature of this part of the hotel, but also the meaning of the word Pimalai, which means 'Heavenly Place' in Thai.

Location Koh Lanta, Krabi, Thailand
Date 2002 (Phase I), 2009 (Phase II)
Interior designer P49 Deesign & Associates
Landscape designer Inside Out Design

MASTER PLAN
1 Beachfront lobby
2 Hillside lobby
3 Beachfront restaurant
4 Hill-top restaurant
5 Spa
0 5 20 50 100m
N
1
2
3
4
5

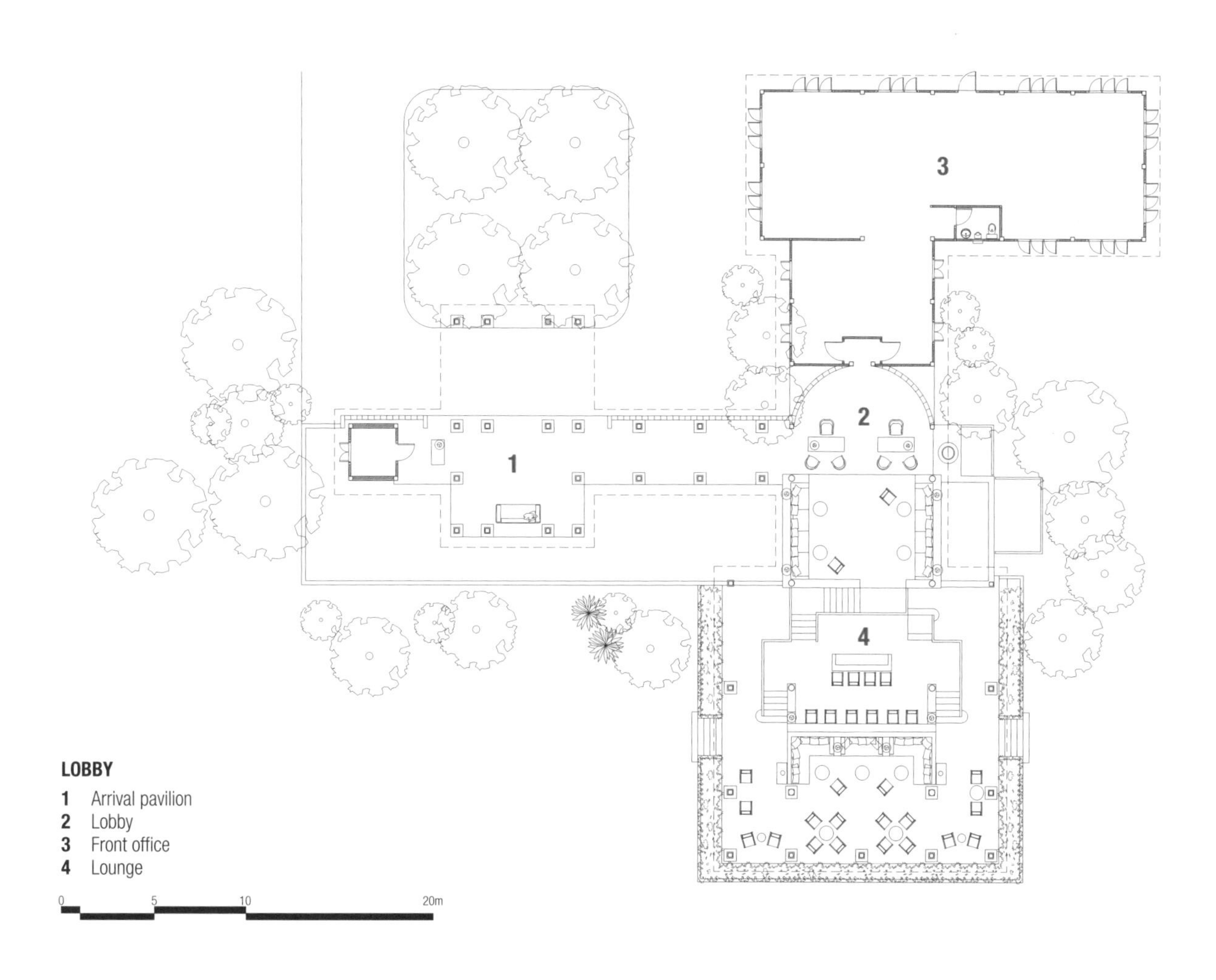

1
2
3
4
LOBBY
1 Arrival pavilion
2 Lobby
3 Front office
4 Lounge
0
5
10
20m

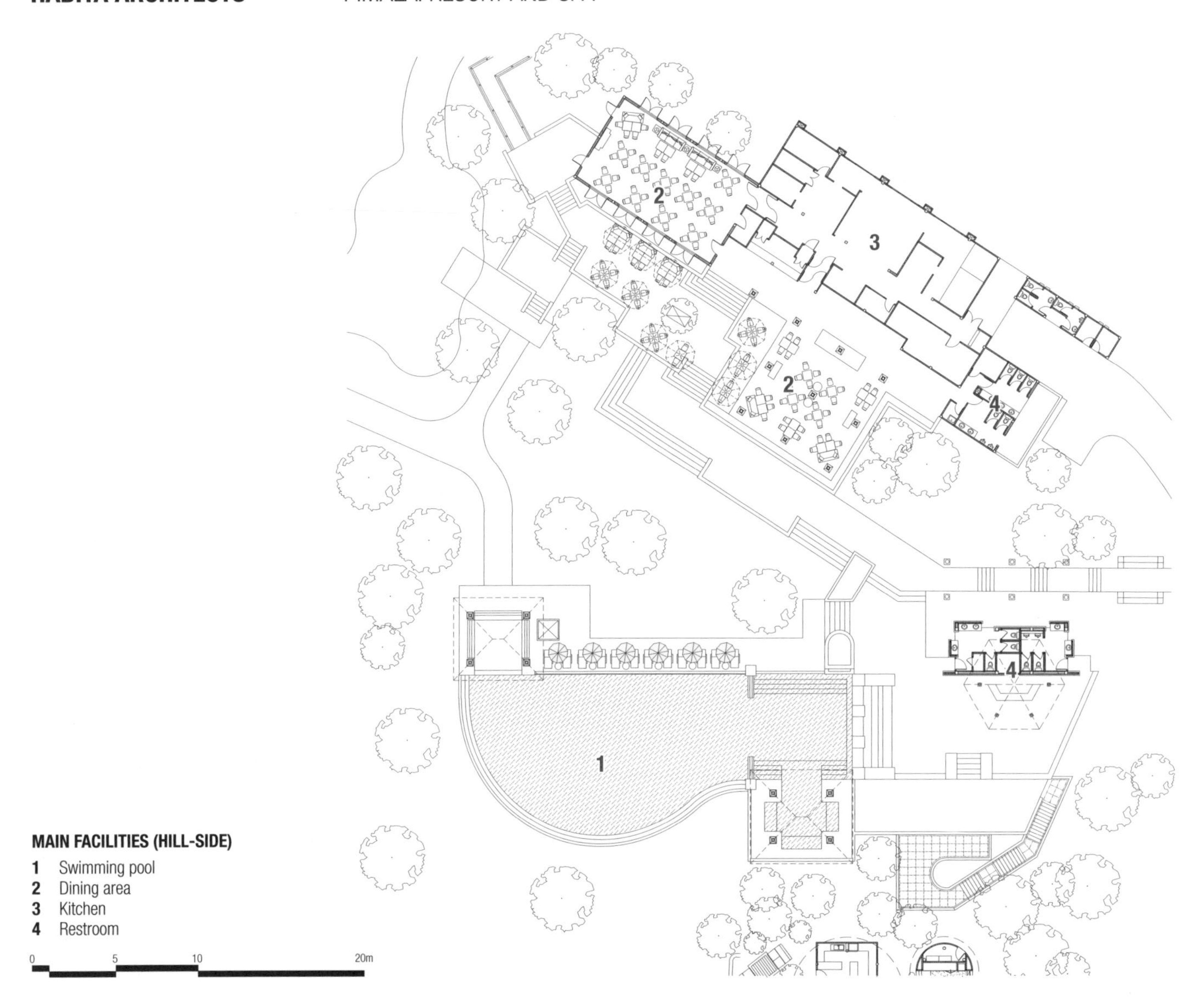
MAIN FACILITIES (HILL-SIDE)
1 Swimming pool
2 Dining area
3 Kitchen
4 Restroom
0
5
10
20m

VILLA

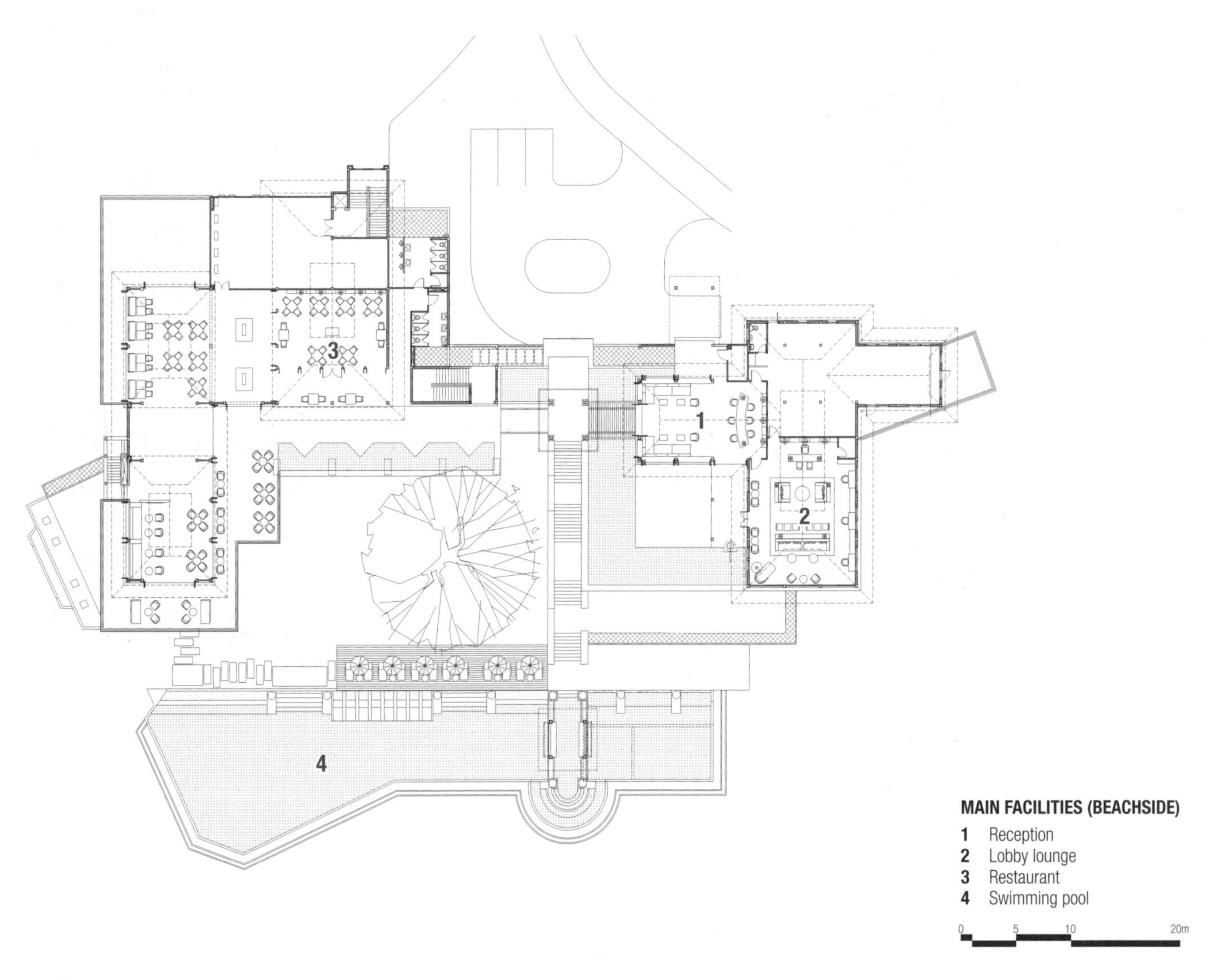
3
1
2
4
MAIN FACILITIES (BEACHSIDE)
1 Reception
2 Lobby lounge
3 Restaurant
4 Swimming pool
0 5 10 20m

苏梅岛布里拉沙乡村酒店
BURI RASA VILLAGE SAMUI

Location Chaweng Beach, Surathani, Thailand
Date 2006
Interior designer AKA
Landscape designer Inside Out Design

Buri Rasa Village Samui is where southern Thai-style village atmosphere meets the most desirable beach on Koh Samui.

The main challenge of this project was to realize the accommodations density required by the client on this narrow beachfront plot, and achieve this within Samui's stringent zoning restrictions.

The guest experience of the southern-style village begins as they enter the resort from Samui's bustling main street. At the front of the resort we placed of a pair of two-story houses, faithfully detailed in the traditional southern Thai style. In traditional style, guests enter by stairs leading to a deck on the upper level connecting the two houses. Screened from the street by large frangipani trees, the upper level comprises a series of serene spaces—the reception occupying the upper level of one house, a small restaurant open to outside guests the other. The dark tone of the timber ceilings and floors are instantly cooling and calming.

We achieved the required guestroom density (there are 32 units in total) by clustering the units together, in the manner of a traditional village. Again, the local convention of entering onto a common deck space is utilized for the upper-level guestrooms. Rooms at ground level are accessed via private gardens, some with plunge pool. The guestroom clusters contain four or eight units, arranged amid coconut palms and other tropical trees, on both side of the twisting village pathway linking reception to the beach.

Materials commonly found in southern Thai architecture, such as natural timber, plastered walls, cement roofing tiles, and double-hip roofs (*langka panya* in Thai) are used.

The village finally opens out to Chaweng beach—characterized by white sand, clear water, and cool sea breezes. Beachfront guest facilities located here comprise restaurant, bar, spa, and swimming pool. Each building carefully emulates traditional southern-style architecture in terms of roof form and detail. Spatially these buildings afford a transition zone between the relative closeness of the guestrooms and the expanse of beach and aquamarine horizon. This is a breathing space where guests can enjoy the sea breeze, and lounge beneath the coconut palms lining the beach.

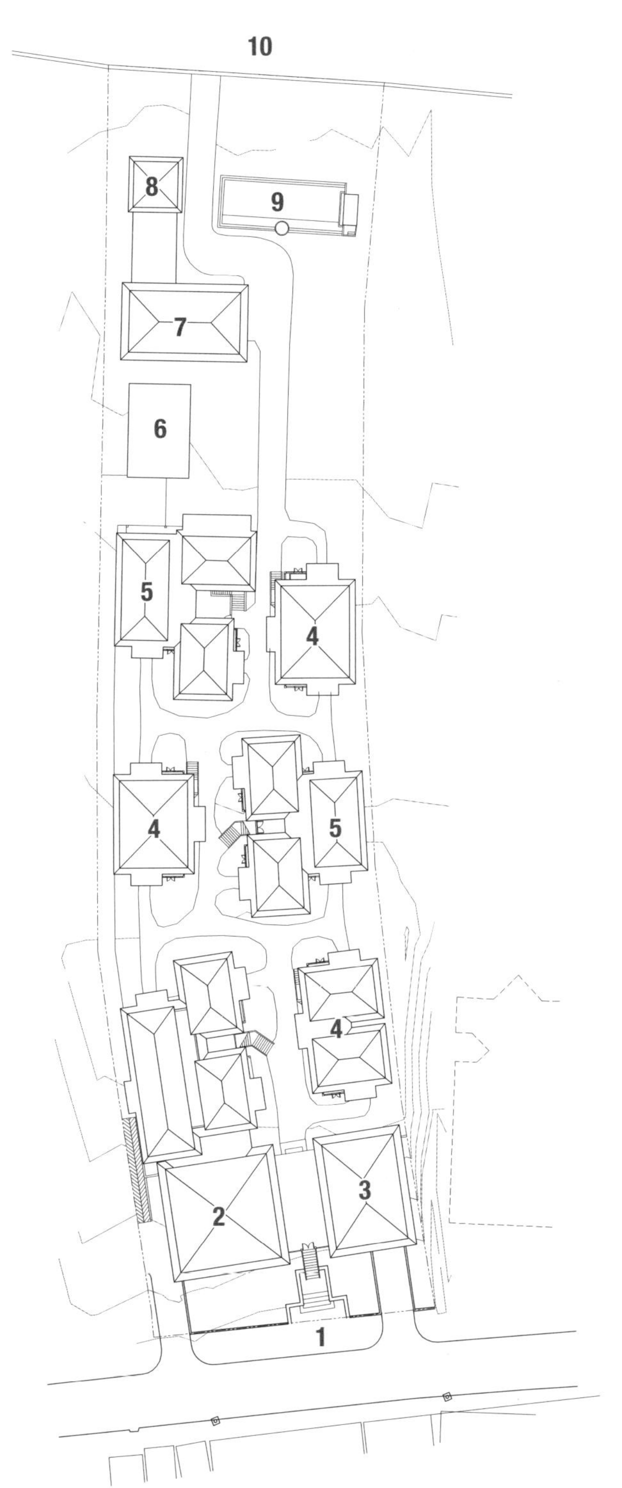

MASTER PLAN

1 Entrance
2 Reception
3 Lounge
4 Guest accommodations (four keys)
5 Guest accommodations (eight keys)
6 Kitchen
7 Restaurant
8 Bar
9 Pool
10 Beach

N

Guest accommodations (eight keys) front elevation

Guest accommodations (eight keys) side elevation

素可泰遗产度假酒店
SUKHOTHAI HERITAGE RESORT

Location Sukhothai, Thailand
Date 2007
Interior designer Son Design by Varavarna Na Ayudhaya

Sukhothai means 'Dawn of Happiness'—traditionally the foundation of the Sukhothai kingdom is considered as the beginning of the Thai nation. This location is therefore of great importance to the Thai people. The international architectural and cultural significance of the Sukhothai Kingdom is also reflected in the single UNESCO World Heritage Site that comprises Sukhothai and two neighboring cities: Si Satchanalai and Kamphaeng Phet.

The Sukhothai Heritage Resort is located near Sukhothai airport on the outskirts of Sukhothai, between Sukhothai Historical Park and Si Satchanalai. The Sukhothai Historical Park is a separate entity to the modern-day city of Sukhothai, and contains all the most significant structures of the former capital.

The influence of history, culture, art, and architecture of the historical park formed the design concept of the resort. Reflecting ponds, grass lawns, and brick structures are the key elements that combine to give the park its uniquely serene ambience. The layout of the resort reconfigures these key elements to reflect the ancient style in modern way.

The Sukhothai Heritage resort comprises 68 boutique rooms and suites and includes two elegant swimming pools, a restaurant, bar, and meeting facilities. These areas are arranged in a very formal overall plan, reflecting the symmetrical arrangement of the main monuments in the park.

After checking in at the open-air lobby, guests enter to central water garden court. This is a broad and tranquil space, occupied by two large pavilions, housed under lofty Thai-style roofs with a gentle inflection. The restaurant and meeting rooms are located on the other side of the large pond, opposite the lobby.

Guests then follow a cloistered promenade to the two-story guestroom buildings, located symmetrically on either side of central court. The two elegant swimming pools at the center of each wing again emphasize the symmetry of the overall arrangement.

In contrast to the central main facilities, the guestrooms have lower pitched roofs and emphasize horizontality. This Wrightian language and orientation of the pools help emphasize the expansion of space from the center outwards.

The prevalence of still water, soft antiquated clay brick, and wide lawns lend a peaceful atmosphere to the entire resort. Guests can enjoy breakfast, lunch and dining around the water garden. Or laze happily in the library or on the private balcony daybed outside their room—from dawn until sunset.

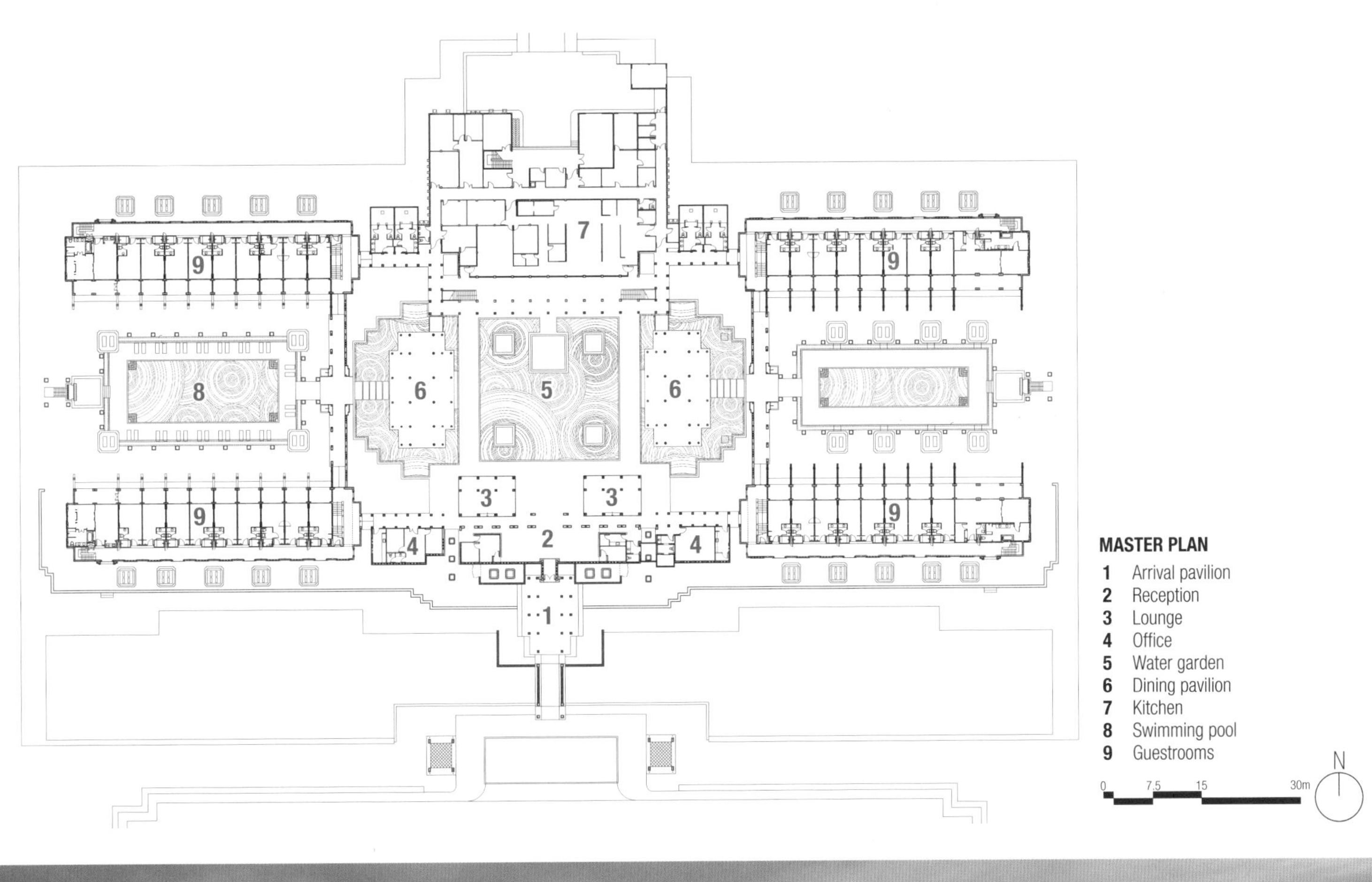

MASTER PLAN

1 Arrival pavilion
2 Reception
3 Lounge
4 Office
5 Water garden
6 Dining pavilion
7 Kitchen
8 Swimming pool
9 Guestrooms

FRONT ELEVATION

0 2.5 5 10 20m

普吉岛斯攀瓦酒店
SRI PANWA PHUKET

The 40-acre (16-hectare) site of this resort occupies the entire Sri Panwa peninsula, located on Phuket's southeastern coast.

The tropical contemporary architectural style allowed us to assimilate three important sources of the unique Phuket culture—southern Thai, Chinese, and Portuguese.

The roofs of the resort's villas are the hip-roof typical of southern Thai architecture, with a deep eaves overhang to deal with the region's six-month-long rainy season. The eaves are raised slightly along their perimeter to create a typical 'Thai' look and express a sense of tenderness. The timber shingles provide a soft color and texture, in harmony with the natural surroundings.

The unique deep orange color theme of the project is derived from local laterite stone. Initially, laterite was mixed into the cement plaster, which created an interesting natural color. Unfortunately, this became difficult to manage in large quantities and to maintain quality control. Therefore, the subsequent orange color is painted on. Despite this, you still feel that these buildings have sprung up from the ground and are component parts of the natural surroundings. In addition, an organic texture is created by sweeping the plaster surface with a hard broomstick.

Other than laterite orange, a Mediterranean blue color is applied to the spa area, the so-called 'Cool Spa.' This color represents sea, nature, coolness, and relaxation. Reflecting the manual nature of spa therapy, the wall texture in the spa is created by hand plastering. Thatch roofing accentuates this handmade concept.

Sino-Portuguese tile was applied in certain areas by the interior designer to reflect this aspect of Phuket's cultural heritage. This integration of cultural references is seen again in the multi-colored timber entry gate to the spa, which is perforated in a star pattern. Upon entering this gate, the guest's view is of a large serene pond with only a few small buildings in sight. The pond is the roof of the large three-story spa building below. The retention of existing trees results in the roof pond adopting free-form lines, in harmony with the ocean and islands beyond.

(See also The Habita @Sri Panwa, page 224.)

Location Cape Panwa, Phuket, Thailand
Date 2008
Interior designer ISM Interior Architecture Workshop; Chemistri Design
Landscape designer Charu—Bhakara; Inside Out Design

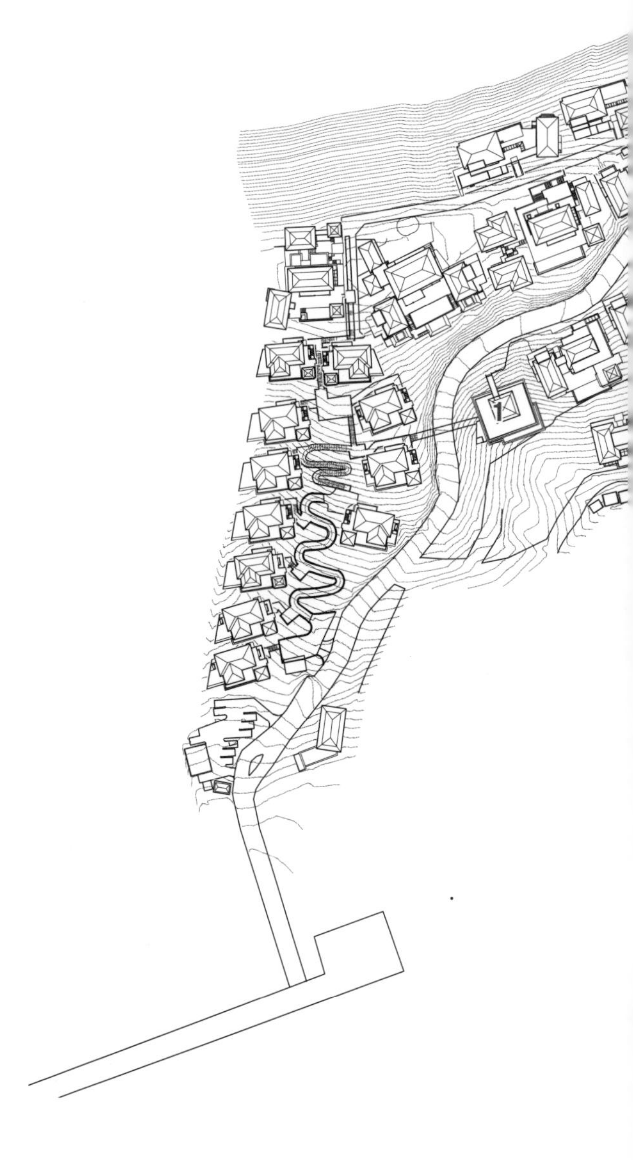

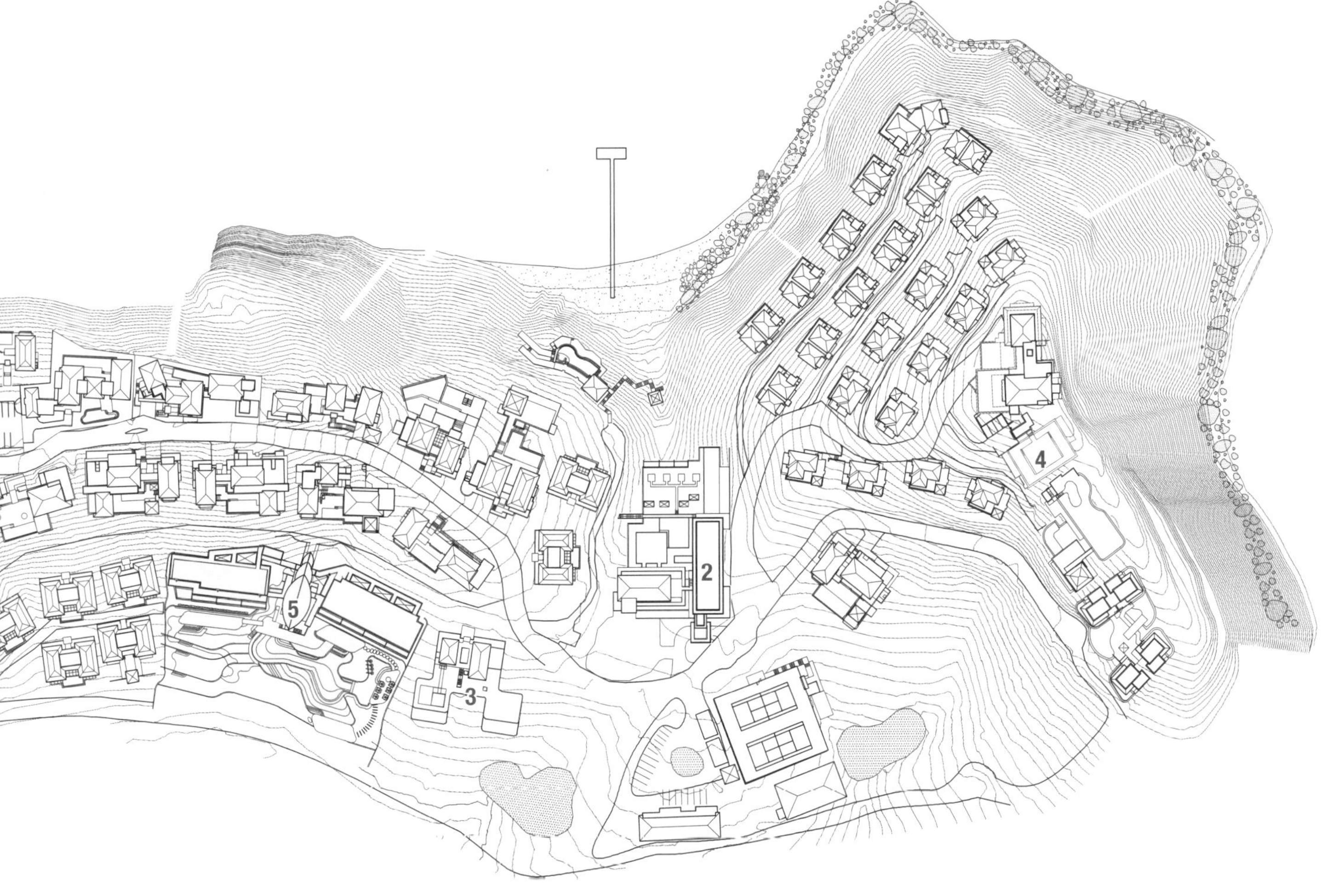

MASTER PLAN

1 Lobby
2 Pool Villa
3 Cool Spa
4 Baba Nest
5 The Habita

0 5 25 50 100m

N

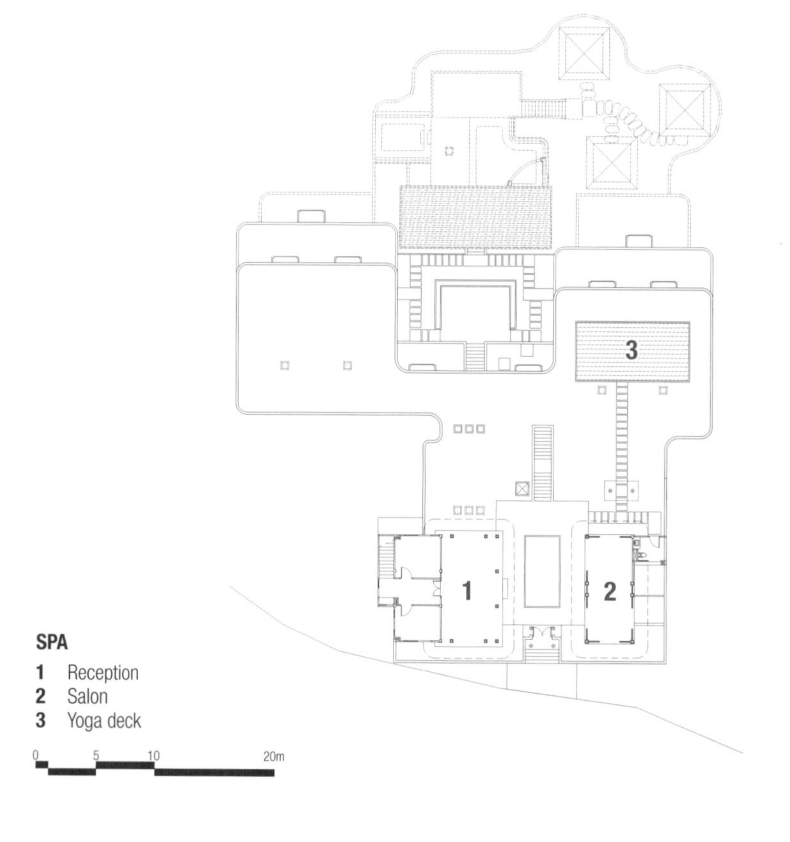
1
2
3
SPA
1 Reception
2 Salon
3 Yoga deck
0 5 10 20m

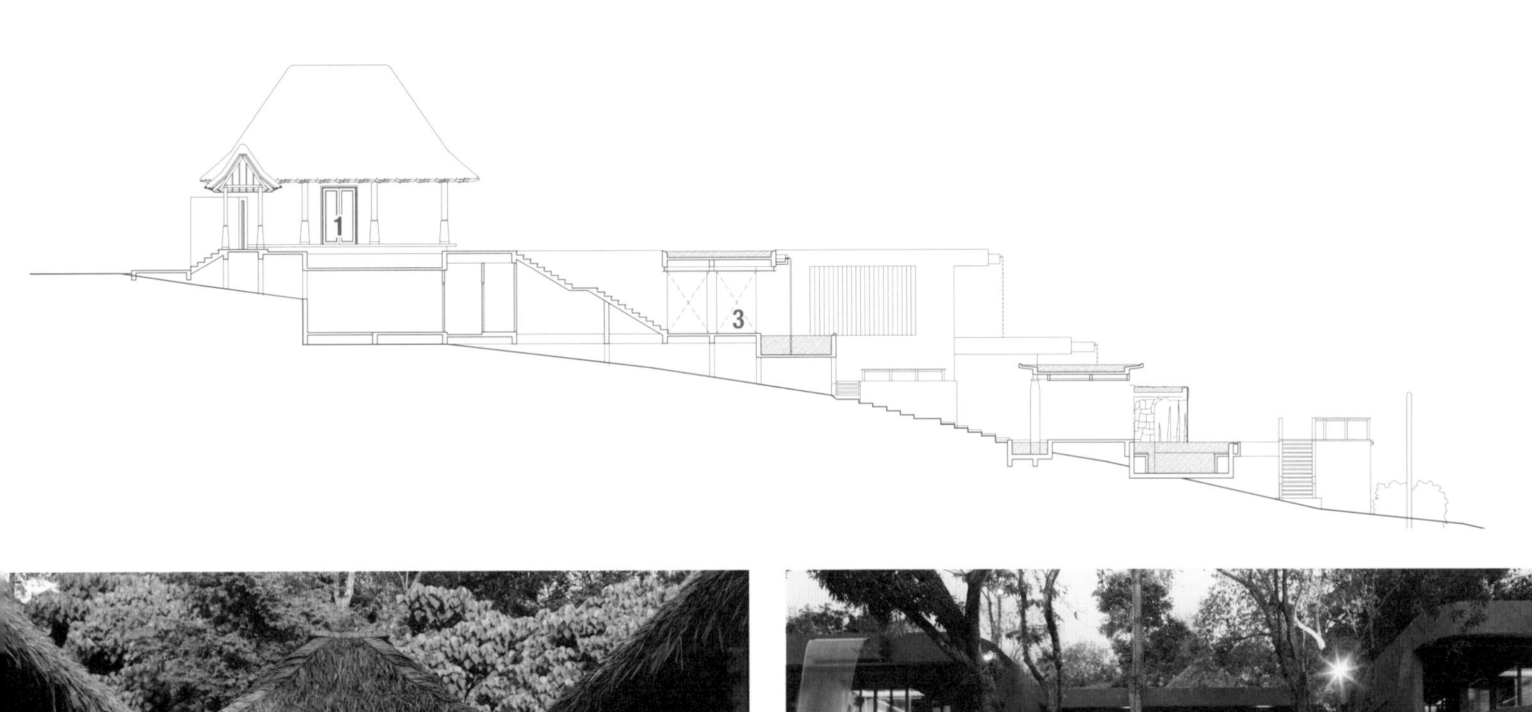
1
3

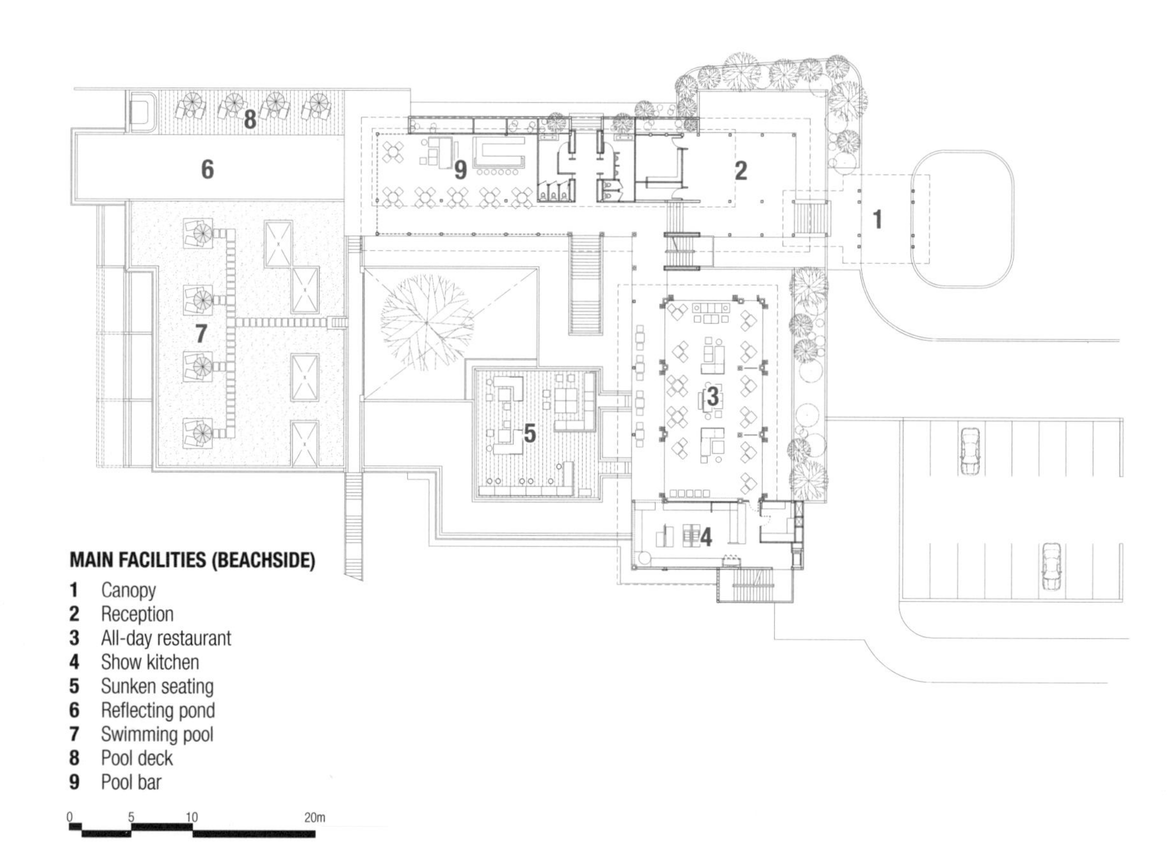

MAIN FACILITIES (BEACHSIDE)

1 Canopy
2 Reception
3 All-day restaurant
4 Show kitchen
5 Sunken seating
6 Reflecting pond
7 Swimming pool
8 Pool deck
9 Pool bar

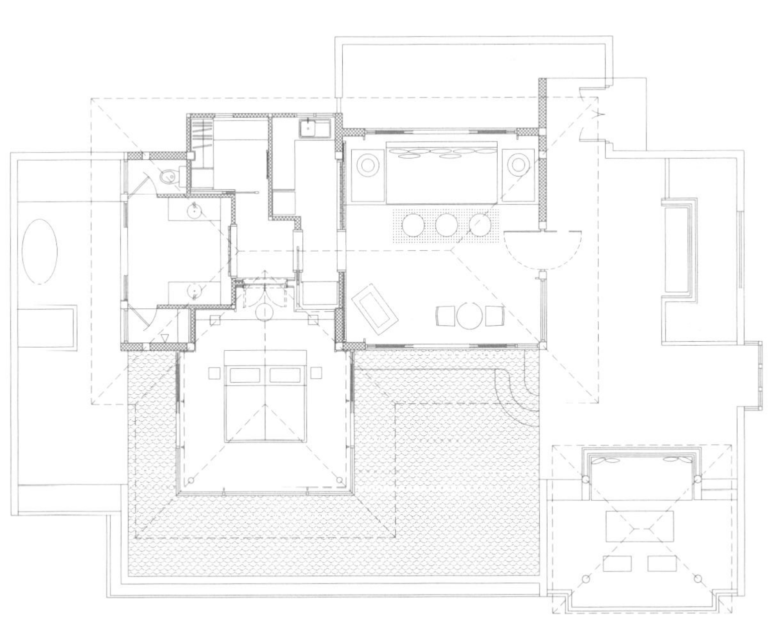
VILLA
0
2.5
5
10m

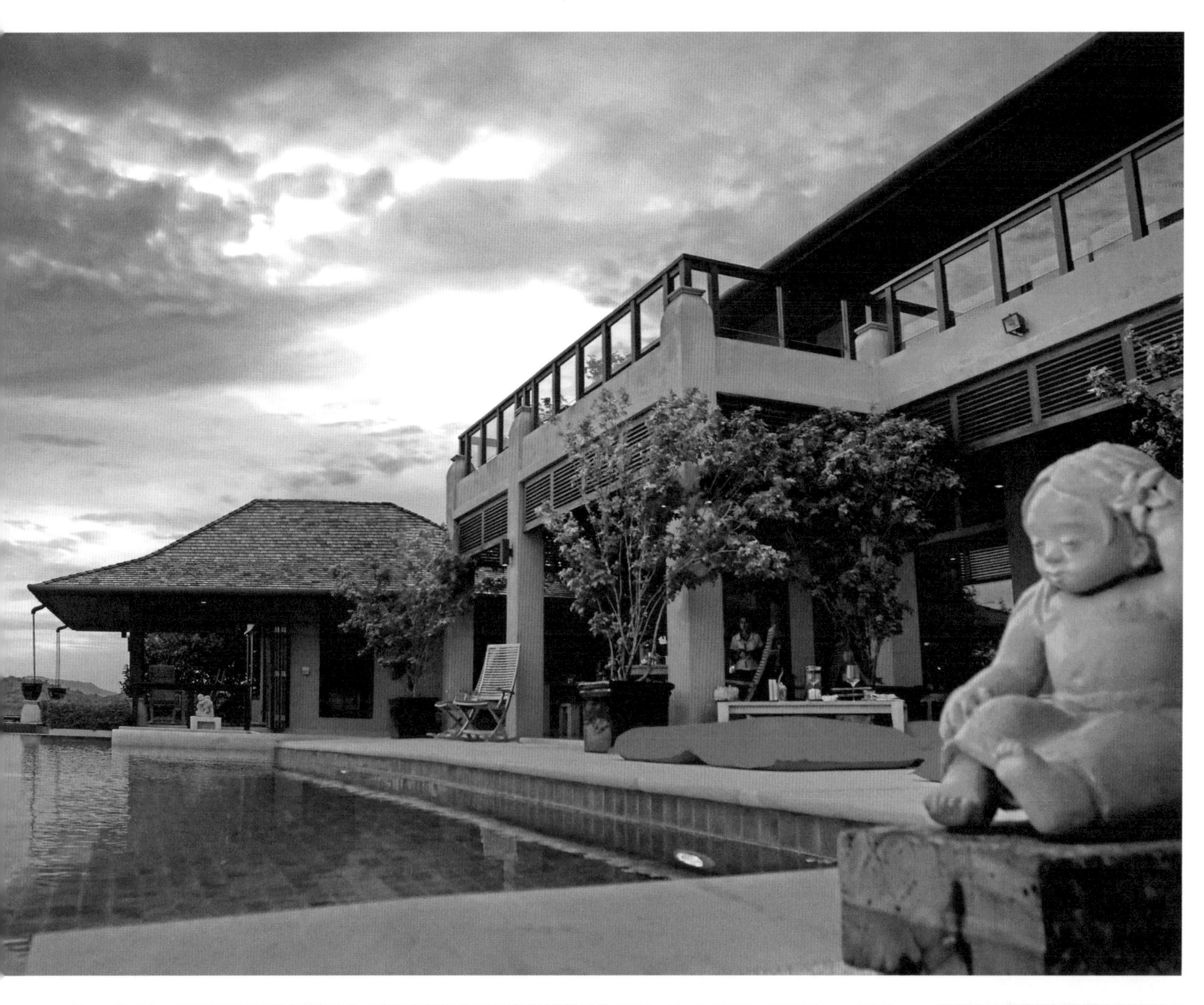

亚雅度假酒店
YAIYA RESORT

Location Cha Am, Thailand
Date 2008
Interior designer Habita Architects
Landscape designer Inside Out Design

Since ancient times, Thai kings and aristocrats have escaped Bangkok's heat for the cool sea breezes of Huahin. The Thai term 'Taak Arkas' means 'to be exposed to the breeze' and expresses the notion of holiday. Unfortunately, the prevalence of air conditioning in the modern world has undermined this habit. So, the design of this beachfront hotel was a concerted effort to re-establish this ambience.

Yaiya Resort, currently known as The Palayana Hua Hin, comprises three zones: the beach club, the 11 pool villas, and the main building, which houses 29 guestrooms, bringing the total accommodations to 40 units.

The beach club is an active area and consists of open-air restaurant, pool, pool bar, and spa. The beachfront is kept as natural as possible and the sand allowed to extend into this area.

The two-storied pool villas are tightly clustered together, yet the upper-level pools maintain individual privacy. The only area where air conditioning is absolutely required—the bedroom—is located at ground level within private gardens. Guests then ascend to the pool deck level. Here the living sala avails itself of cool breezes and a tranquil outlook over the pool and garden.

The main building contains the lobby and guestrooms, which are arranged over five floors. This gives the main building a mass appropriate to the depth of the site. Guests arrive in the open-air lobby at upper ground level, the slope of the site permitting a discrete lower level parking garage. The staff and back-of-house facilities are located adjacent to the lobby, in a walled compound.

Guest circulation is a meandering axis from the main building down to the beach, mimicking the gentle slope of the site. The landscaping comprises simple lawn, and large existing tropical trees and shrubs.

Rather than copying traditional Thai architecture, the design of both the villas and main building are simple, unpretentious, and modern, with some local touches. The main hotel block has a rational façade, the deep verandas of each guestroom softened with bamboo blinds. The pool villas are mainly plaster, with limited elements of natural timber, including the shingle roofs, which have acquired a softness in the salt air.

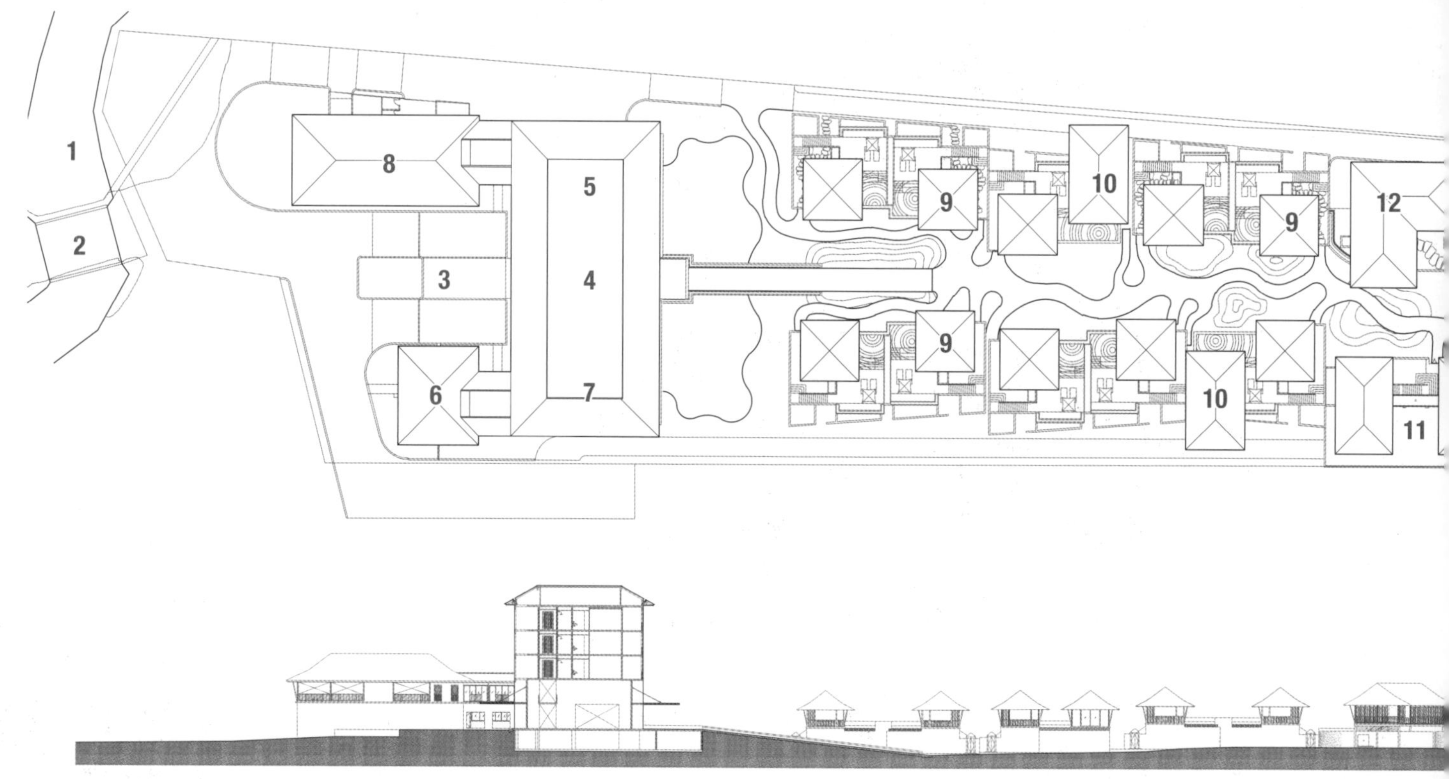
1
2
3
4
5
6
7
8
9
9
9
9
10
10
11
12

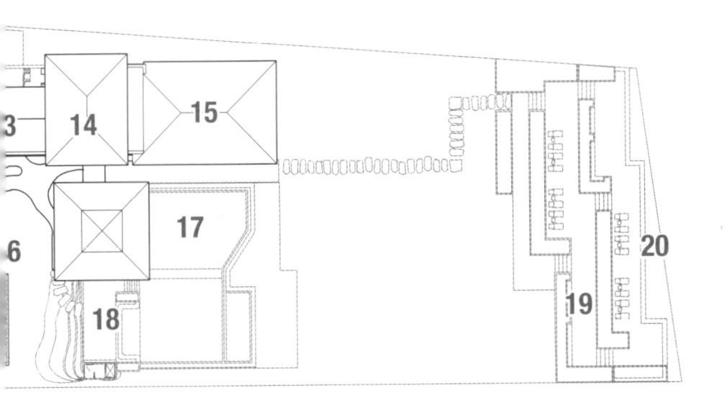

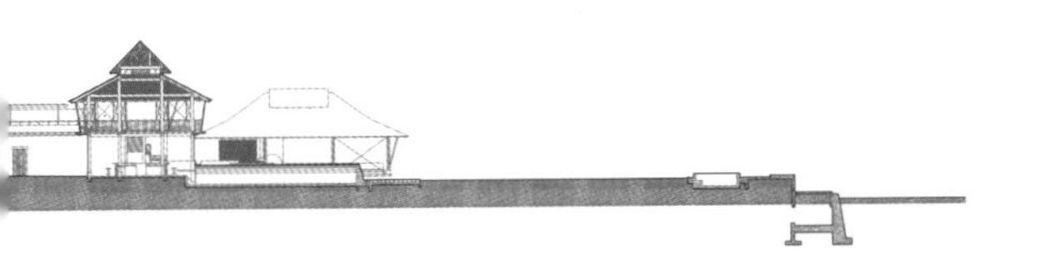

MASTER PLAN

1 Creek
2 Bridge
3 Drop-off
4 Gallery
5 Reception
6 Retail
7 Guestroom
8 Office & back-of-house
9 One-bedroom pool villa
10 Two-bedroom pool villa
11 Three-bedroom pool villa
12 Spa
13 WC
14 Display kitchen
15 Restaurant
16 Pool bar
17 Main pool
18 Deck
19 Step terrace
20 Beach

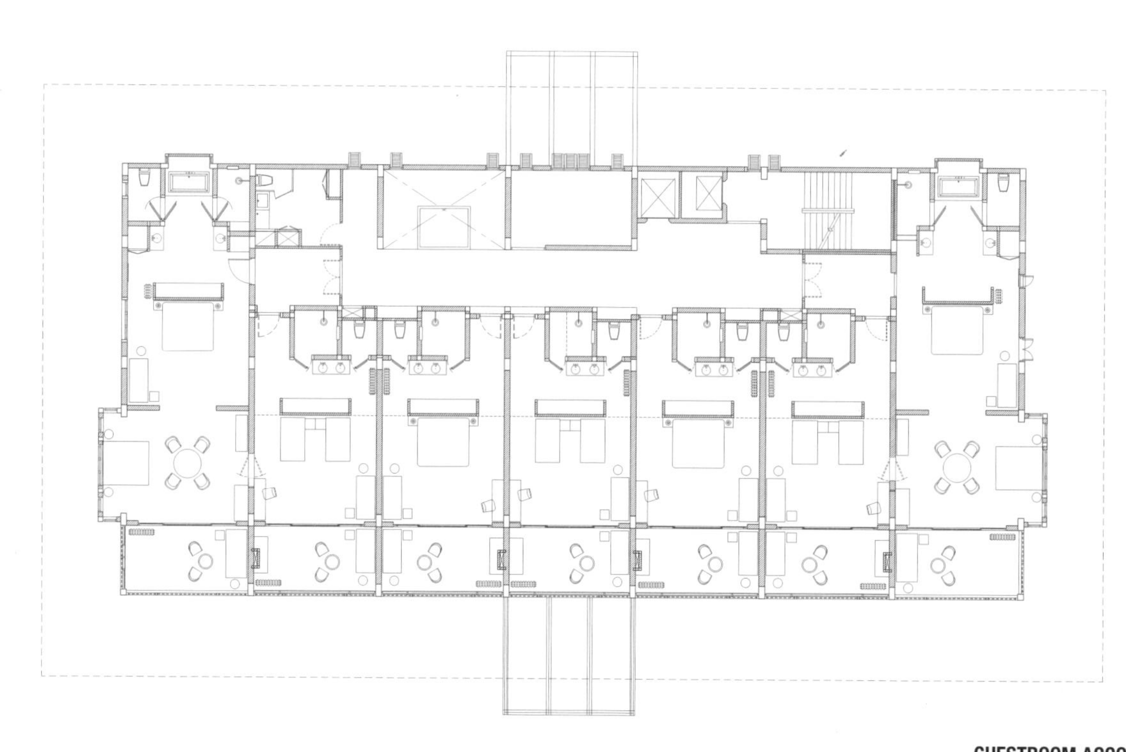

GUESTROOM ACCOMMODATIONS

ONE-BEDROOM VILLA

1 Bedroom
2 Entrance
3 Bathroom
4 Garden

0 1 5 10 20m

137柱子之家酒店
137 PILLARS HOUSE

Location Chiang Mai, Thailand
Date 2011
Interior designer P49 Deesign & Associates
Landscape designer P Landscape

137 Pillars House was originally the northern headquarters of the East Borneo Trading Company, built 125 years ago in the center of Chiang Mai, Thailand. At the onset of the project, the once-elegant teak building (known also as Baan Borneo) was in a dilapidated state and liable to flood in the wet season.

As well as saving and restoring Baan Borneo, the challenge was also to create a hotel environment that stimulated the imagination of guests and moved them back in time, but did not just blindly replicate the old house.

While the old house retains its stately presence, the new buildings are clearly identified with the use of contrasting materials. The entrance pavilion, reception sala and spa are in elegant white-painted timber. These structures are clustered at the street entrance to the hotel. From this area, guests meander under century-old trees towards the old house, which gradually reveals itself across a wide lawn. Along the route stand clusters of new two-storied villas. The ground-floor levels are distinguished from the old house by a weighty white plaster. The upper stories are made of timber and detailed in a similar fashion to Baan Borneo. The 30 suites all have cement tile roofs, while the original structure has a timber shingle roof.

Another challenge was how to deal with the annual flooding of the site. Although raised off the ground by around 3 feet (almost 1 meter), flood waters would still come up to Baan Borneo's floorboards. The decision was made to lift the entire house and the lawn by approximately 5 feet (1.5 meters). This maintained the proportional relationship of the original house with the ground. It also allowed an additional lower floor to be inserted behind the original house for the hotel's back-of-house facilities.

The original house's functions are now limited now to a lavish library and lounge. These spaces connect to the restaurant to the rear, which occupies original outbuildings, now raised to sit over the back-of-house facilities.

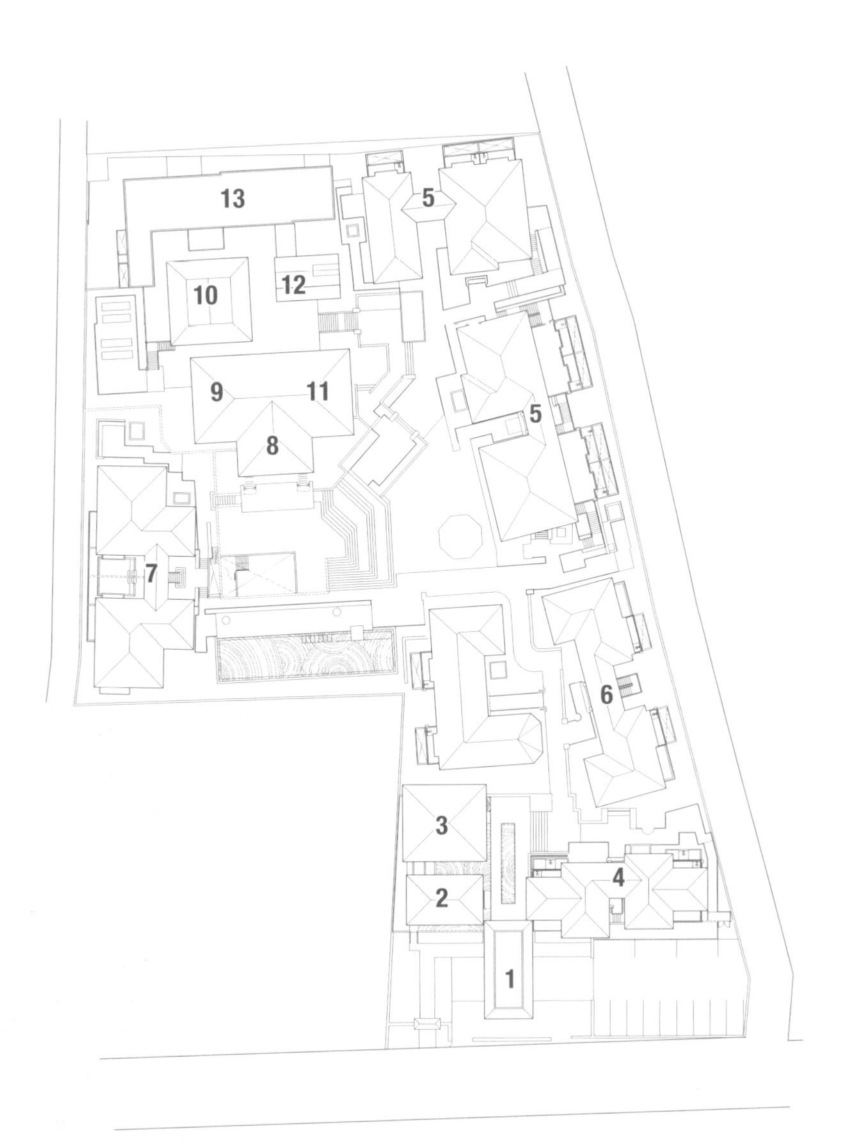

MASTER PLAN

1 Entrance
2 Lobby
3 Spa
4 Standard guestroom (King—Type A)
5 Standard guestroom (King—Type B)
6 Standard guestroom (Twin)
7 Suite Villa
8 Lounge
9 Library
10 Restaurant
11 Meeting room
12 Open kitchen
13 Kitchen

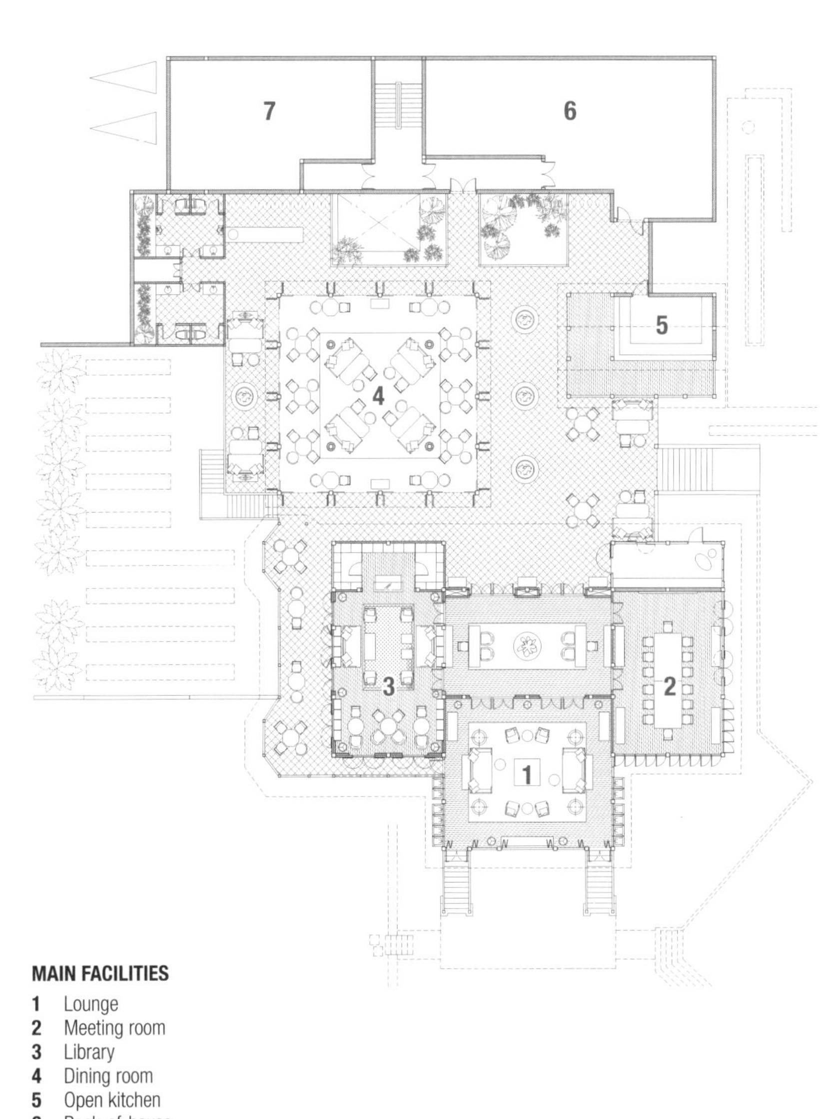

MAIN FACILITIES

1 Lounge
2 Meeting room
3 Library
4 Dining room
5 Open kitchen
6 Back-of-house
7 Kitchen

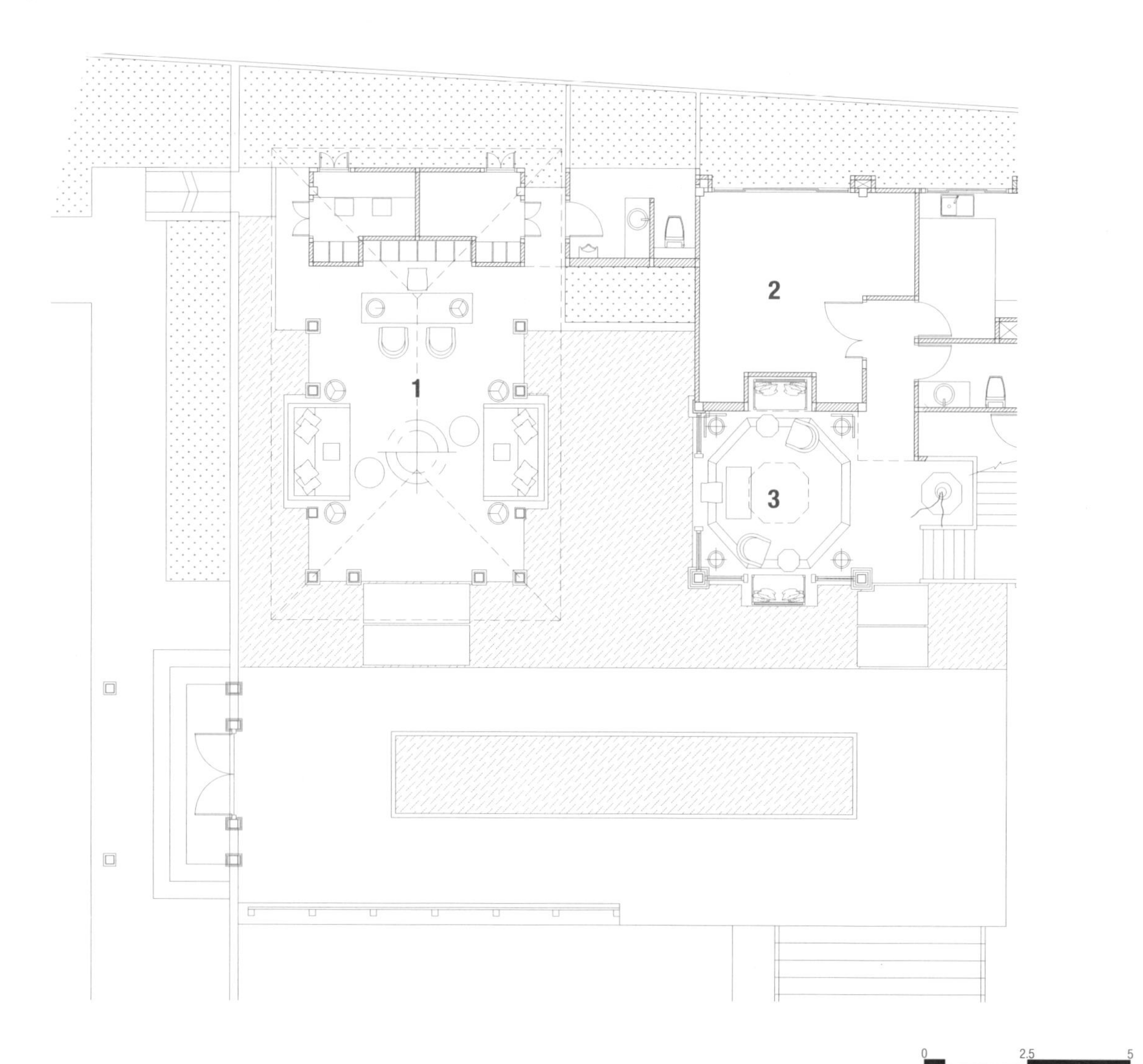

RECEPTION & SPA

1 Reception
2 Spa reception
3 Spa service

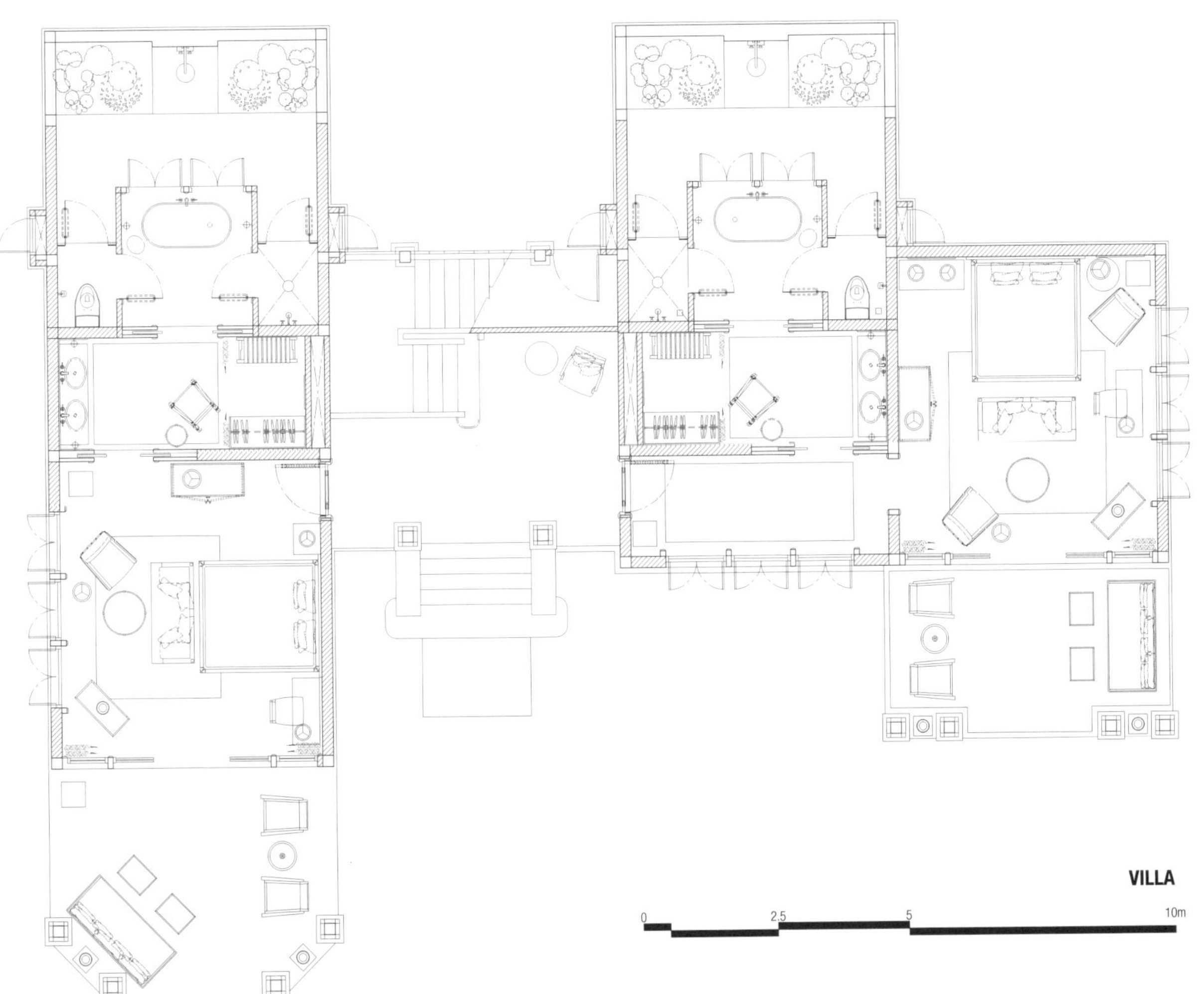
VILLA
0
2.5
5
10m

普吉岛迪奈涵酒店
THE NAI HARN

Location Phuket, Thailand
Date 2016
Interior designer P49 Deesign & Associates
Landscape designer Inside Out Design

Opened in 1986 as the only luxury resort on the island of Phuket, the Nai Harn (formerly the Royal Phuket Yacht Club) remains today the only member of 'Leading Hotels of the World' located in this most famous of all Thailand's beach resorts.

After 30 years, its spectacular location notwithstanding, the 110-unit resort required renovation to meet changing guest expectations.

Habita's scope of works firstly focused on re-conceptualizing public spaces, including restaurant, wellness center, outdoor amenities, and main reception lobby. These areas comprise a three-story plinth positioned on the rocks overlooking Nai Harn beach.

Secondly, there was an opportunity to refine the modernist multi-terraced guestrooms, arrayed on the hillside behind the public areas. After reconfiguring some of the original suites, the renovation works resulted in a resort of 130 units in total.

The major structural intervention involved removing the traditional gabled roofs of the main restaurant and guestroom lift tower. This presented the Nai Harn to the bay with a new clarity and elegance—and provided space to create a new rooftop deck experience. Intimate seating areas are sunken into a reflecting pool, the edge of which merges into the panoramic view over the Andaman Sea. Opening onto the roof deck is a new rooftop bar and wellness center comprising gym and spa.

A cooling color palette of white with strong accent color was introduced to the guestroom interiors and private terraces. Perforated concrete block and metal screens visually soften key front-of-house areas, such as the reception and the guestroom access stair tower.

The overall result is simplified Modernist timelessness. This befits both the esteemed status of the Nai Harn, as well as its prize location. The rediscovered Modernist rhythm of the resort harmonizes with the deep tropical forest hillscape in which it is set, while contrasting with the azure waters below. Across the bay lies Promthep Cape, the southernmost point of Phuket island—a timeless location to view the setting sun descend into the Andaman Sea from the private terraces, restaurants, and outdoor areas of the Nai Harn.

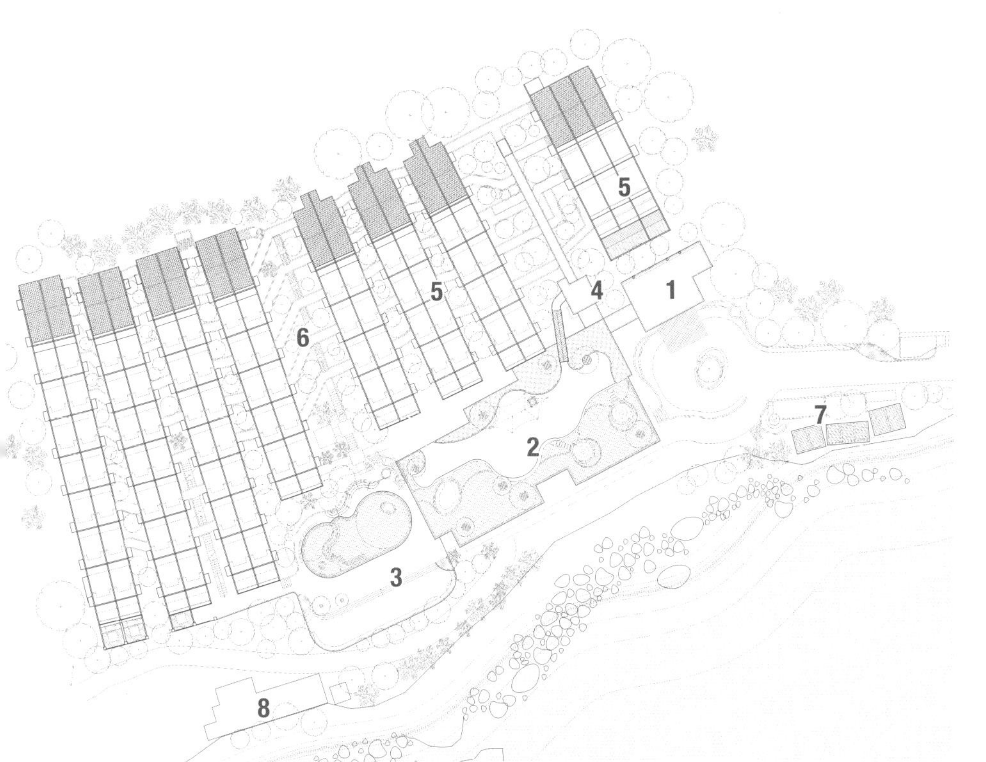

MASTER PLAN

1 Reception lobby
2 Main building & Reflections rooftop bar
3 Swimming pool
4 Stair & lift tower
5 Guestroom block
6 Central stair
7 Beach restaurant
8 Engineering shed

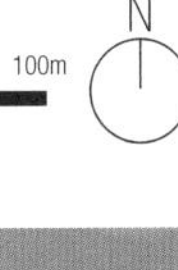

MASTER PLAN

1 Reception lobby
2 Main building & Reflections rooftop bar
3 Swimming pool
4 Stair & lift tower
5 Guestroom block
6 Central stair
7 Beach restaurant
8 Engineering shed

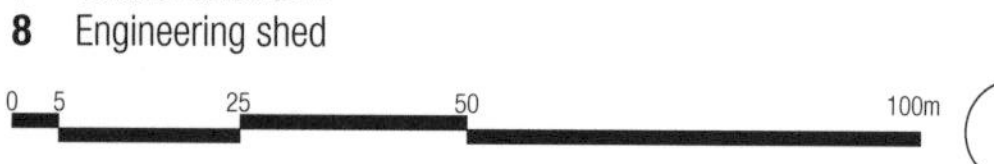

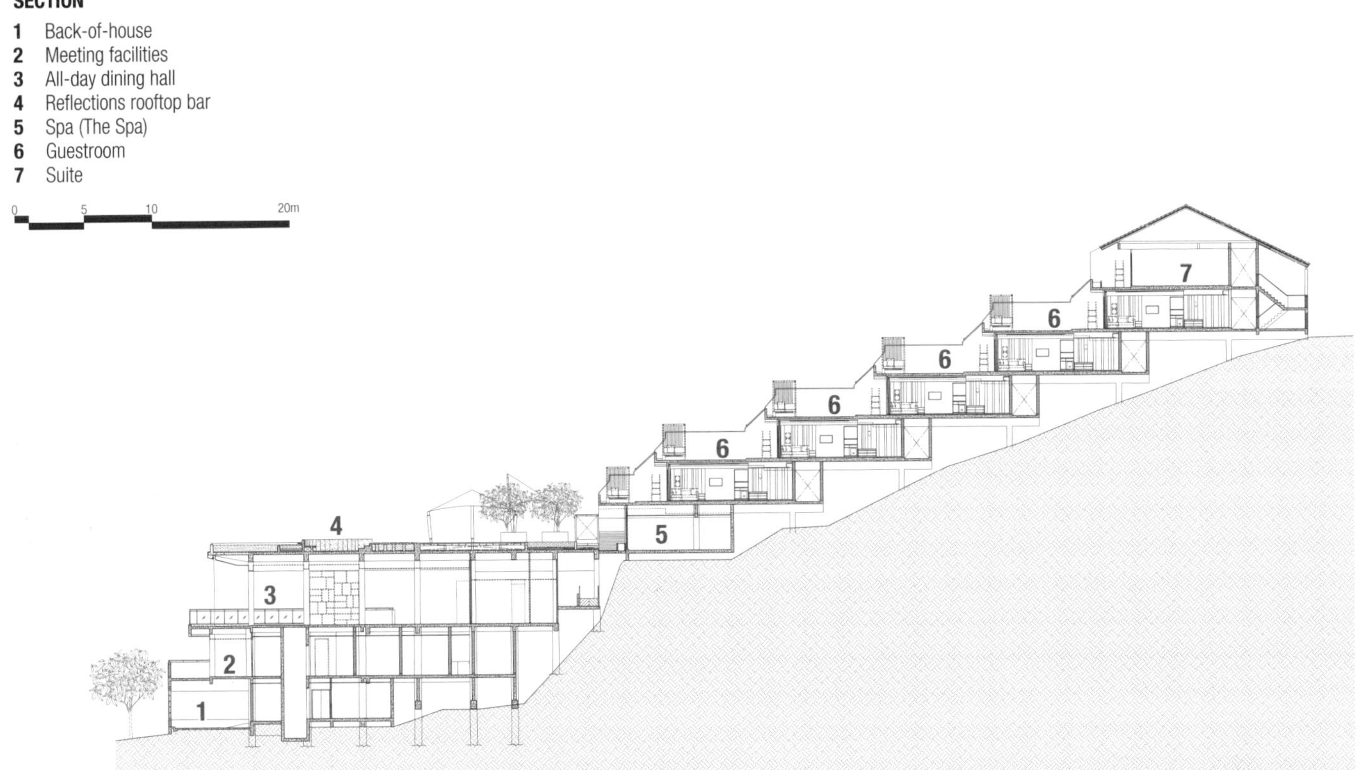
SECTION
1 Back-of-house
2 Meeting facilities
3 All-day dining hall
4 Reflections rooftop bar
5 Spa (The Spa)
6 Guestroom
7 Suite
0
5
10
20m
1
2
3
4
5
6
6
6
6
6
7

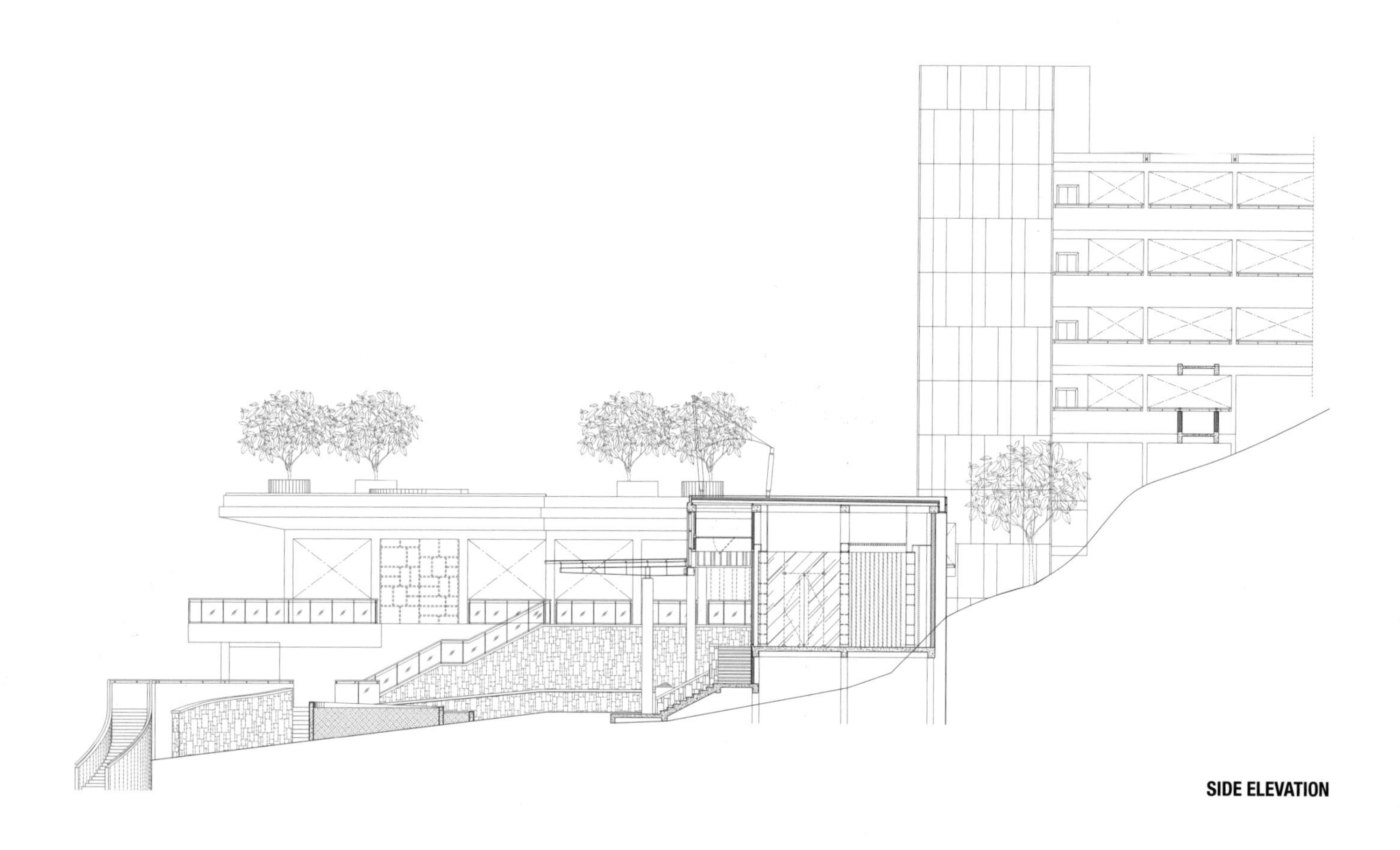

SIDE ELEVATION

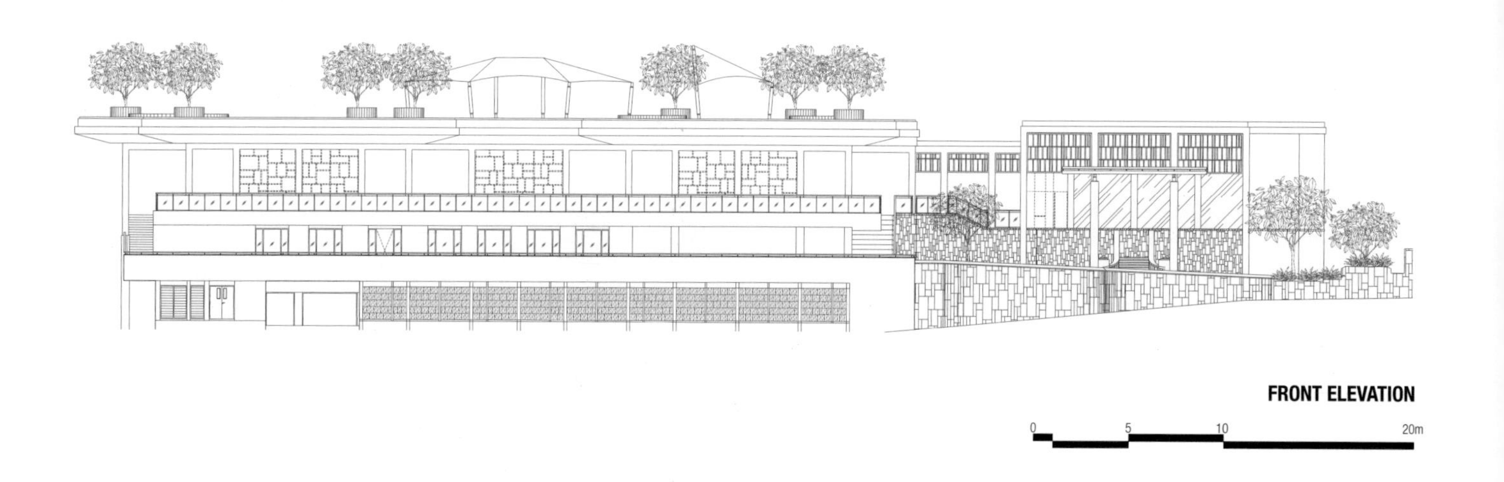

FRONT ELEVATION

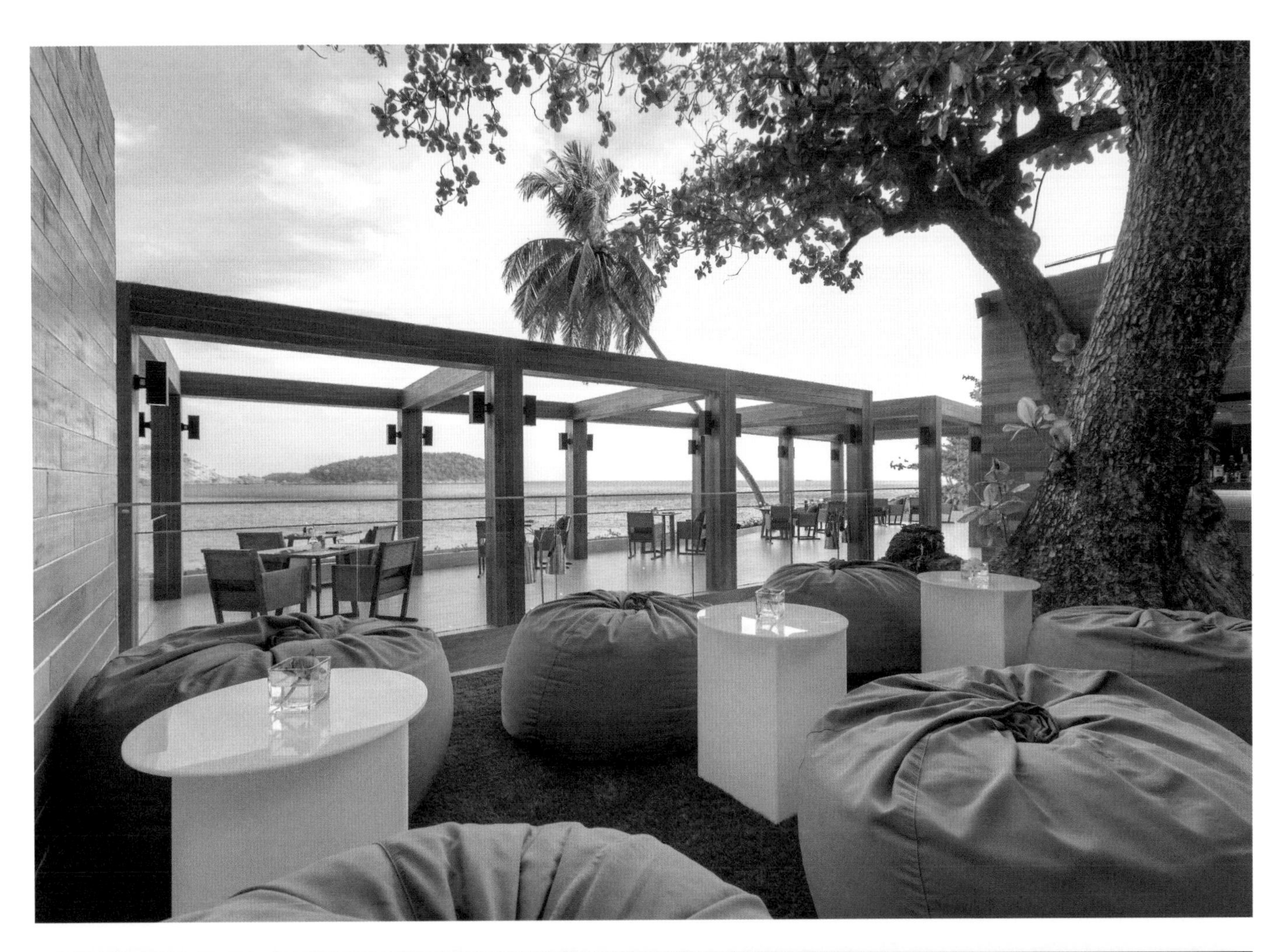

普吉岛斯攀瓦酒店

THE HABITA @SRI PANWA

The Habita is a part of the Sri Panwa hotel, also designed by Habita (see also Sri Panwa Phuket, page 186). Whereas the existing Sri Panwa resort solely comprises individual pool villas, the Habita provides an alternative arrangement of 30 rooms, each with private pool. Each pool is located to provide unobstructed views from Cape Panwa over the Gulf of Thailand.

The site was a green area where many large tropical trees were located. Given the density of development required, it was unavoidable that some of these trees. The project concept became focused on how to compensate for the lost trees. The project aims to partially restore the equilibrium by incorporating meaningful tree-like forms into the architecture, as well as providing plenty of space for natural sea breezes to circulate around these forms.

The outstanding architectural feature of the project is the free-formed roof that bisects the body of the hotel, running from the road entry all the way to the pool. Its form is directly inspired by tree leaves, with the ceiling pattern under the roof reflecting the veins of the leaf.

The second strong feature incorporating the physical nature of trees is the masonry screen wall. The precast concrete blocks are inspired by the shape of tree branches. These screens are inserted into the hotel rooms, as if the rooms are enclosed with nature, while also providing the rooms with privacy.

Tropical vernacular architecture is also presented in the design, supporting the more overt references to nature. For example, in the order of columns that demarcate the approach to the hotel and the interface between column and free-from leaf roof.

Considerable attention is given to the sequence of space and the various interactions of guests and their surroundings in different areas of the project. For example, how the sea and pool view are perceived initially from the drop-off, then from the reception, the lounge, the restaurant, or even from the room corridor. These are essential parts of the overall hotel experience for the guest.

Location Cape Panwa, Phuket, Thailand
Date 2016
Interior designer Chemistri
Landscape designer P Landscape

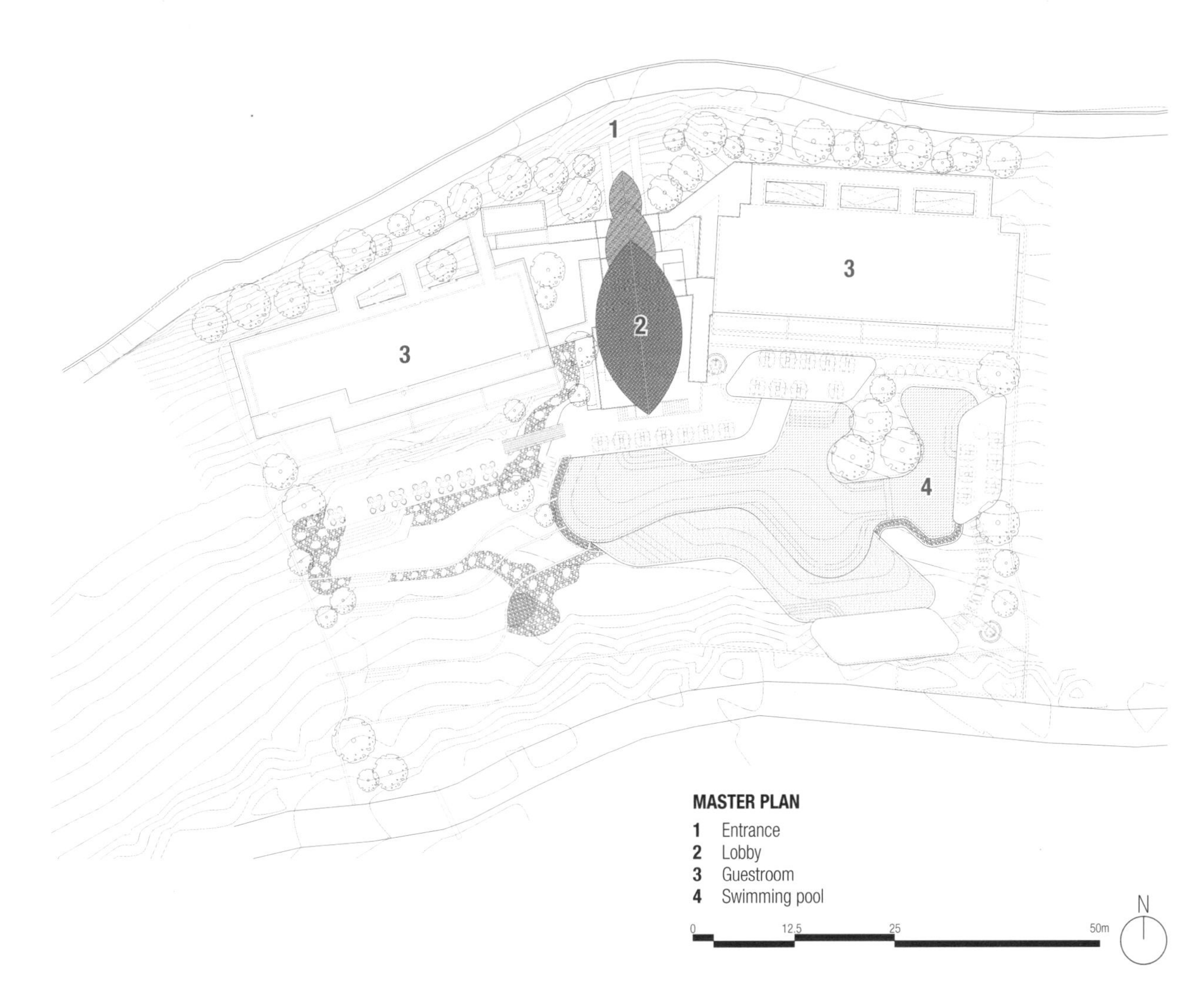

MASTER PLAN

1 Entrance
2 Lobby
3 Guestroom
4 Swimming pool

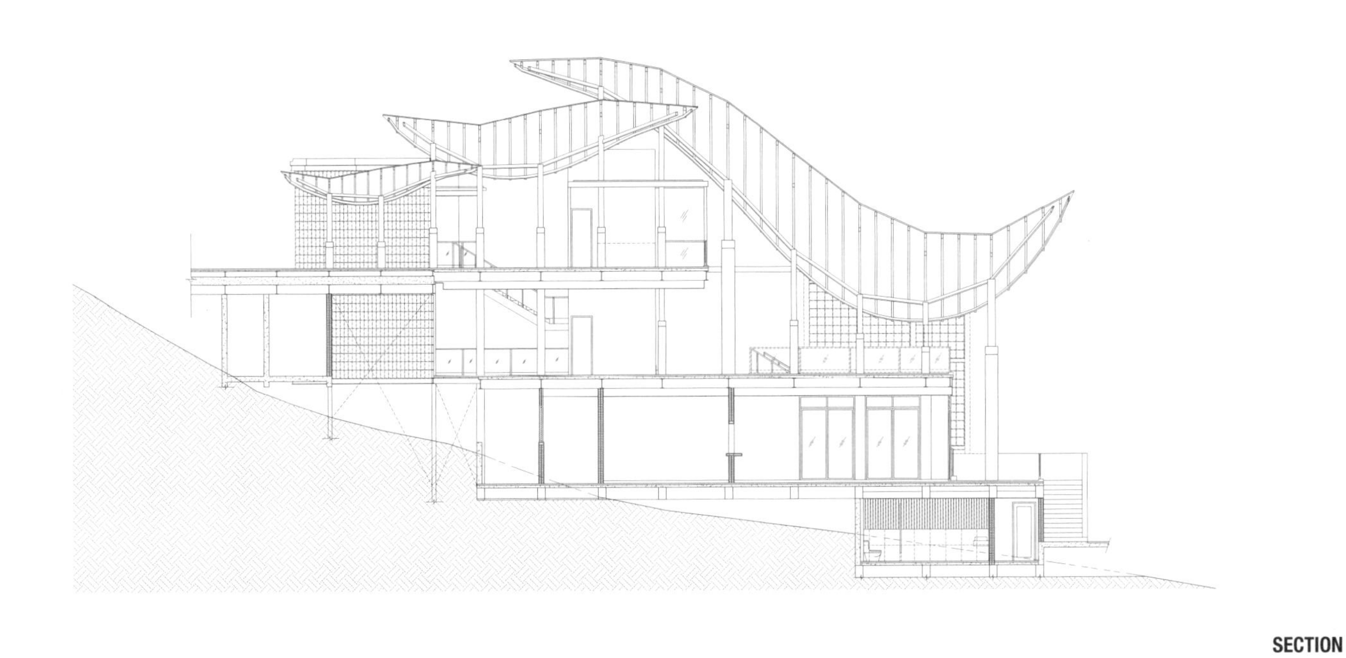

SECTION

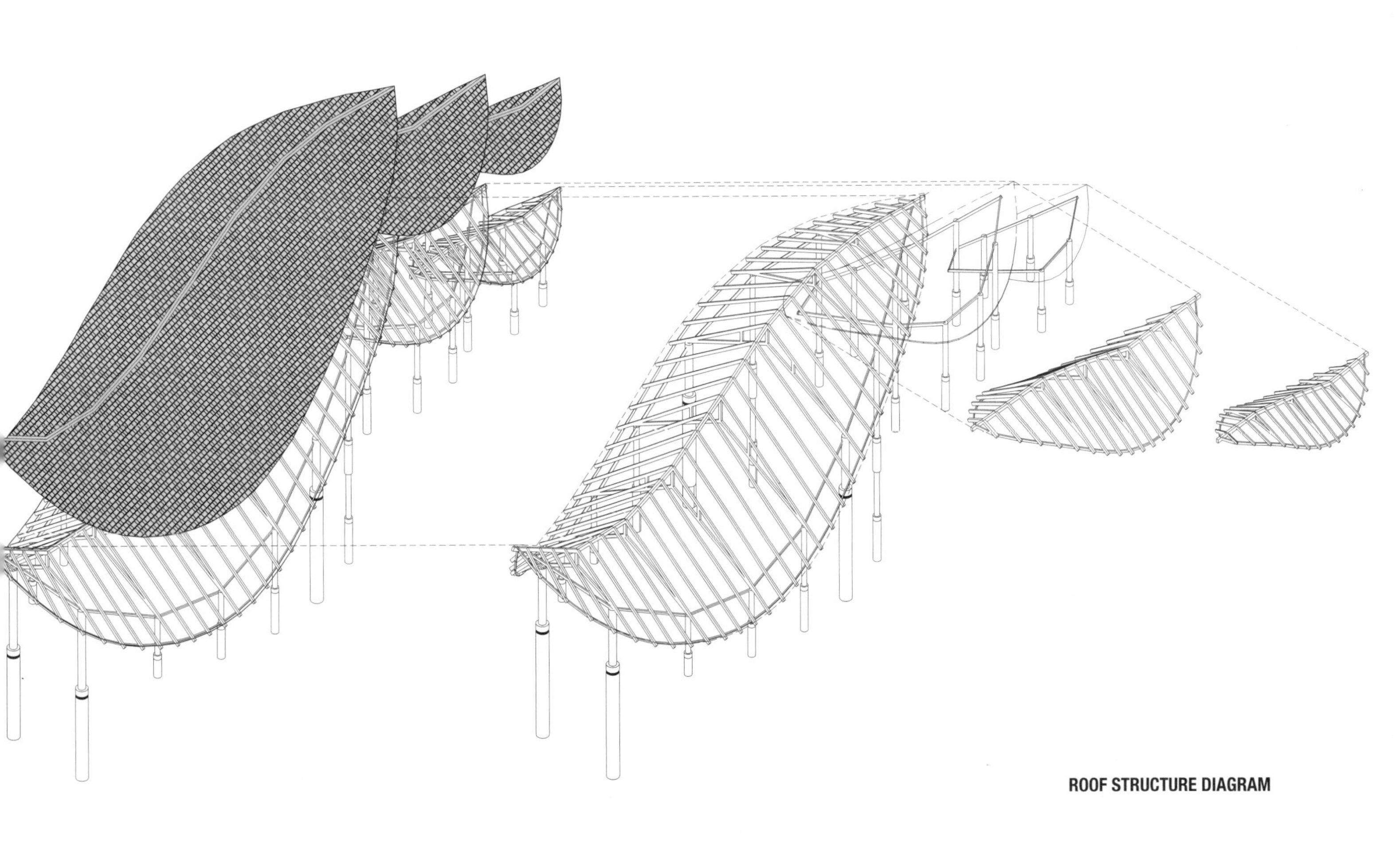

ROOF STRUCTURE DIAGRAM

在建项目

ONGOING PROJECTS

不丹六善酒店
SIX SENSES BHUTAN

Six Senses Bhutan presented a unique and challenging project in terms of size and location. The hotel operates differently from other resorts, by being separated into smaller hotels across a number of locations. Each of these five small hotels is located at a major tourist—Thimphu, Paro, Punakha, Gangtey, and Bumthang. Put together, the hotel comprises a total of 82 units.

The challenge was to adapt each hotel to its physical surroundings and also attain individuality for each hotel concept. For example, the renderings shown (page 238) depict the concept for Punakha. Here rice fields are a predominant geographical feature and are incorporated into the design.

The traditional Bhutanese house is an outstanding example of sustainable architecture. Building materials consist of timber, stone, and earth—all of which are abundantly available locally and can be sustainably harvested. The rammed earth walls provide excellent insulation and provide interior spaces with a constant temperature. In the summer when the outside temperature goes up, the rooms become cooler. Conversely, when the outside temperature goes down in the winter, the rooms become warmer.

Location (Thimphu, Paro, Punakha, Gangtey, Bumthang), Bhutan
Date Under construction
Interior designer Six Senses Creative Department
Landscape designer P Landscape (for Thimphu & Paro site)

BUMTHANG

PUNAKHA

THIMPHU

GANGTEY

PARO

华欣巴巴海滩度假村
BABA BEACH CLUB

Baba Beach Club is a luxury residential and vacation villa estate and beachfront hotel and beach club, comprising 55 suites and 11 pool villas, located on Cha Am beach. The Modern Colonial style concept for the project is inspired by the palaces and houses built in the area in the past. Cha Am and neighboring Hua Hin have been popular for beach-destination holidays for almost a century. This is evident in the existence of many old palaces and holiday houses, which were constructed in colonial architectural style that was popular at the time. These places have become important tourist destinations due to their unique architectural characteristics. The design concept of Baba Beach Club connects the glory of the colonial style from to the present day.

Location Cha Am beach, Phetchaburi, Thailand
Date Under construction
Interior designer Chemistri
Landscape designer P Landscape

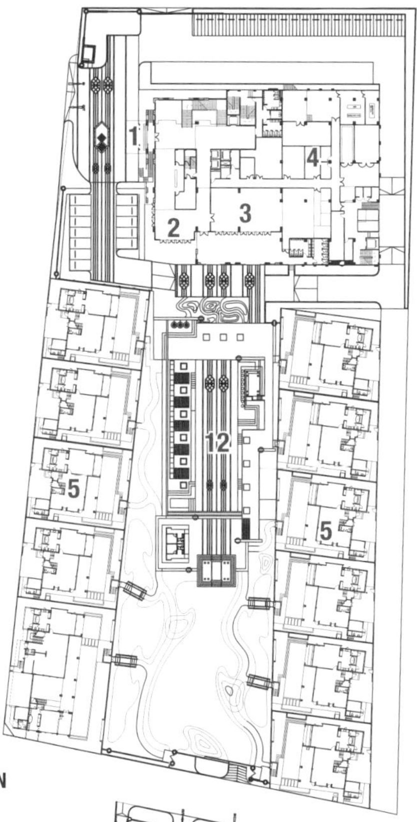

MASTER PLAN

1 Drop-off
2 Lobby & reception
3 Restaurant
4 Back-of-house
5 Hotel Villa
6 Reception
7 Back-of-house
8 Guestroom
9 Restaurant
10 Kitchen
11 Bar
12 Main swimming pool
13 Beach swimming pool

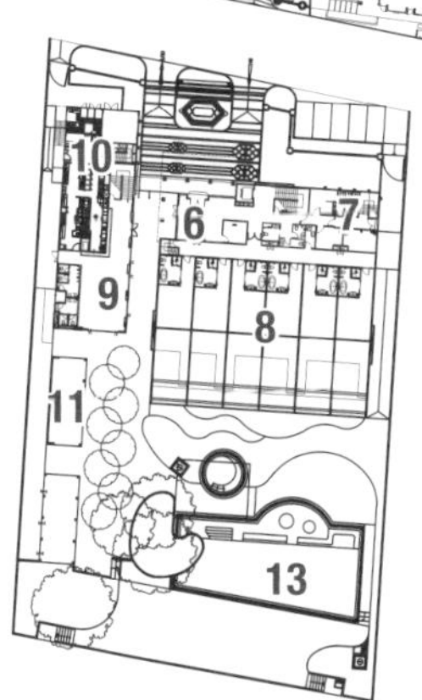

芭东英迪格酒店

HOTEL INDIGO PHUKET PATONG

Hotel Indigo Phuket Patong is a boutique hotel developed for the International Hotel Group (IHG) under the corporate motto 'No two neighborhoods are alike.' The hotel is located at the northern end of Patong Beach, the vibrant nightlife hub of Phuket.

This mid-height resort-styled hotel will provide 180 rooms in total, comprising 10 room types. It will also feature a themed café and bar, boutique retail stores at street level, and two swimming pools, one of which overlooks the ocean.

A principal architectural feature of the hotel will be the 'Blue Gate,' an atrium that divides the street elevation in two. This will be clad in vivid blue porcelain tiles, reflecting the heritage of Phuket as a notable overseas Chinese community and trading port on the Andaman coast. This featured wall is repeated at the second five-story-high atrium entrance to the neighboring café. The atrium void is topped with roof garden, connecting the corridor and the sky.

Guestroom layouts are categorized into two main types, depending on balcony size. The arrangement of the rooms develops a dynamic rhythm along the 330-foot-long (100-meter-long) façade. The larger size of balcony will be provided with oversized seating, wooden drinking hatch, and high stools, to take in the street view. The gesture toward the neighborhood's Chinese heritage is repeated in the Oriental patterns to be incorporated into the balcony privacy screens.

In contrast to the calm, modern exterior, the interior design will pick up on the vibrancy of the surrounding area—reflecting the activity of local fishing villages, the rhythm of the tropical jungle, and the energy of the nearby international nightlife district.

Location Patong Beach, Phuket, Thailand
Date Under construction
Interior designer Blink Design Group
Landscape designer Inside Out Design

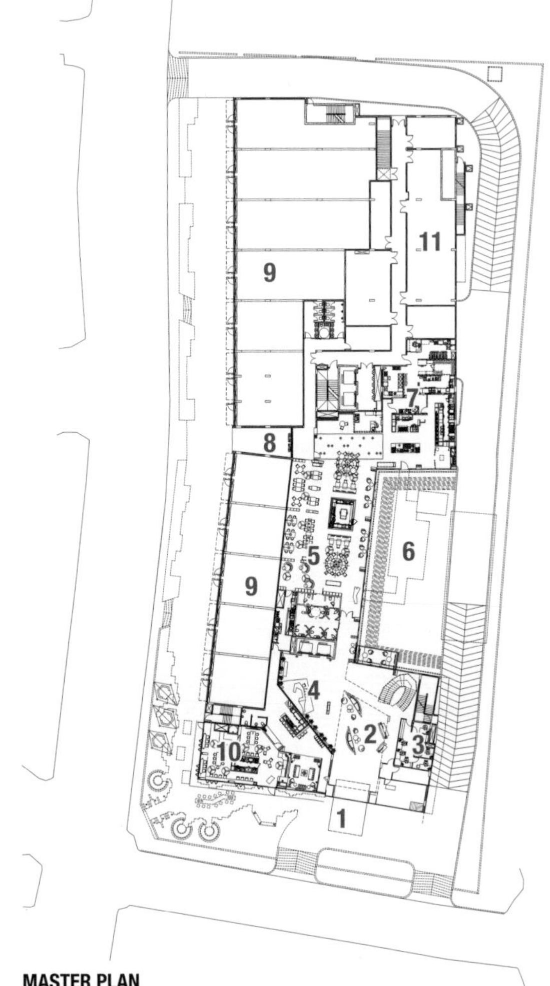

MASTER PLAN

1 Drop-off
2 Lobby & reception
3 Lounge
4 Front office
5 Neighboring café
6 Outdoor café
7 Main kitchen
8 Second entrance
9 Retail
10 Bar
11 Mechanical & electrical

普吉岛卡马拉海滩洲际度假酒店

INTERCONTINENTAL PHUKET RESORT KAMALA BEACH

Location Kamala Beach, Phuket,Thailand
Date Under construction
Interior designer P49 Deesign & Associates
Landscape designer Belt Collins Thailand

This resort of 222 rooms and villas is located on Kamala Beach, on the western coast of Phuket. It occupies two distinct land plots. One is next to the beach, where existing pine trees are retained as much as possible. The other makes up the larger portion of the resort and is located on empty flat land next to the hillside.

The design concept was to create on this vacant land a visual and architectural representation of 'Thai-ness.' The hotel became a platform to re-tell the 'Traibhumikatha,' the oldest Thai text, which sets out an explanation of heaven. Thai artists have re-interpreted the Traibhumikatha through painting and sculpture from many perspectives. For this project, the theme of heaven was selected for architectural interpretation, focusing on four central components:

The Form of Heaven: In mural paintings in ancient Thai temples, heaven's pavilions are interpreted to be tall and grand in shape, with multi-roof tiers with the uppermost spire reaching up to the sky. This is still the motif used by Thai artists today. The spire form is applied to the resort's main pavilion, which forms the centerpiece of the development.

The Sky on Earth: The reflective glass mosaic tile is used as the wall finish to external walls of the Cluster Villas, in order to lighten the mass. These walls enclose the axial way to the main pavilion. The reflection of the sky from the mosaic wall will make the guests feel like as if they are walking on the clouds.

The Passage to Heaven: Another perception of heaven is that it exists up above, not on the ground. This is symbolized by the stairs ascending to the main pavilion.

The Angels' Floating Abode: Buildings in heaven are believed to float in the sky and be surrounded by the sacred ocean. In this way, the main pavilion is surrounded by a large lotus pond. The reflection of the sky on the pond and the mosaic wall leading to the main pavilion reinforce this idea.

MASTER PLAN

1 Beachside reception
2 Specialty restaurant
3 Beach bar
4 Guestroom
5 Beach pavilion
6 Main lobby & reception
7 All-day dining
8 Loungc
9 Cluster villa
10 Hotel block
11 Pool villa
12 Spa

N

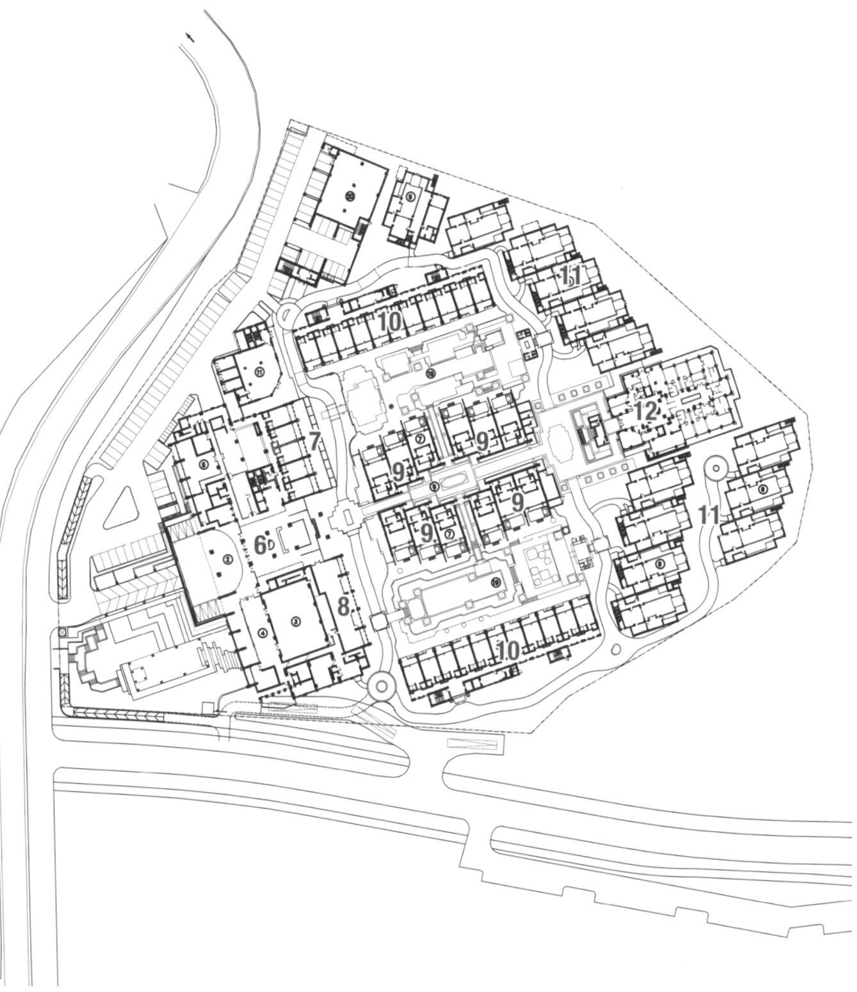

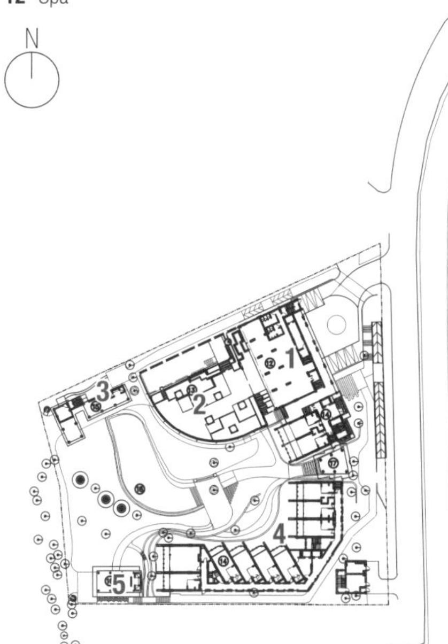

清迈孟寨温泉酒店

ONSEN @ MONCHAM

Our client came to us with a love for the architecture and lifestyle of Japan. Especially the onsen, or traditional Japanese bath house. They required a home for their retirement, located in the mountains where the temperature is cool all year round. The site has a wonderful view up a mountain valley, with strawberry fields growing on the mountainside. The house was to be designed as a retreat for their family and friends, equipped with an onsen bath house and 14 guestrooms for their many visitors.

Habita's design was based conceptually on a small Japanese village. Characteristic of such villages are the different materials and subtle changes of design applied to each house. This responds to specific owner requirements, while retaining a simple method of construction. The buildings are connected by a small pathway, reminiscent of a pathway through a mountain. The main house is situated at the top level to make the most of the panoramic view. The onsen is located at the lowest point of the land for privacy and for a specific view of the mountain. Architecturally, the main feature of the project are the two distinct styles of roof, contrasting black clay tiles and timber shingles, which makes this project a distinctive building in the village.

Location Chaing Mai, Thailand
Date Under construction
Interior designer Design T2
Landscape designer Inside Out Design

MASTER PLAN

1 Drop-off
2 Lobby & reception
3 Coffee shop
4 Restaurant
5 Onsen
6 Guestroom
7 Garden
8 Staff area

项目信息

CREDITS

PHOTOGRAPHY

137 PILLARS HOUSE
AMARIN PRINTING & PUBLISHING PLC
ANANTARA MAI KHAO PHUKET VILLAS
ANANTARA RASA NANDA KOH PHANGAN VILLAS
APIRAK SUKSAI
BURI RASA VILLAGE SAMUI
KANATE CHAINAPONG
LAYAN RESIDENCES BY ANANTARA
LI-ZENN PUBLISHING
MERCURE BANGKOK MAKKASAN
PAKKAWAT PAISITTHAWEE
P LANDSCAPE
PULLMAN PHUKET ARCADIA NAITON BEACH
RUNGKIT CHAROENWAT
SHERATON HUA HIN PRANBURI VILLAS
SITTISAK NAMKHAM
SIX SENSES LAAMU
SIX SENSES NINH VAN BAY
SIX SENSES QING CHENG MOUNTAIN
SIX SENSES DESTINATION SPA PHUKET
SIX SENSES YAO NOI
SONEVA FUSHI
SONEVA JANI
SONEVA KIRI
SOOPAKARN SRISAKUL
SRI PANWA PHUKET
SUKHOTHAI HERITAGE RESORT
THE NAI HARN
WARODOM NIMMANAHAEMINDA
WEERAPON SINGNOI

全体员工
HABITAINS

CORE

KRISDA ROCHANAKORN—PRINCIPAL
PISIT SAYAMPOL—PRINCIPAL
PRAMOT KRABUANRATANA—ASSOCIATE PARTNER
JARAS PONGPIENRAK—ASSOCIATE PARTNER
DIREK WONGPANITKRIT—ASSOCIATE PARTNER

STAFF (2017)

ARCHITECTS
DAMRONG SUBAN
THANASARN TIENTRAKARN
PATIVET CHAIYASOT
PRAPAN WUTTHIWONGYOTIN
RICHARD MINTON MORRIS
PAEN ROCHANAKORN
WITHAWAT LOETPHAISANKUN
VIPAPORN AREEROM
VEERAPONG EAWPANICH
NOPPHARIT SUANPHO
PAKKAWAT PAISITTHAWEE
GRAIWIT DERMLIM
RACHAKORN THANYAMONGKOLCHAI
PATIPAN POTIYARAJ
KANOKPAN BOONKOOM
CHUENCHIT PANICHKUL
WIWAT LERKAMNOUYCHOKE

ADMIN
PICHAYA SRIDAENG—STUDIO MANAGER
SIRANEE PANNAK—ACCOUNTANT
THONGLA CHANTASILA

TECHNICIANS
PONGPETCH PATTAPONG
RAVIPARS BOOTKROOT
TEERASAK CHANGHETPOL

FORMER STAFF

(CONTRIBUTORS TO THE PROJECTS PUBLISHED IN THIS TITLE)

SARAN SOONTORNSUK—ASSOCIATE PARTNER
ATHIWAT CHANTASUWAT
JITINART SIRITED
JITTAPOO SUPA OPAS
KANATE CHAINAPONG
PANYAPON SOPAN
PITIRAT YOSWATTANA
PRAPATWET SUKHO
THANANART KORNMANEEROJ
NAREERAT KLINHOM
SIRICHAI MOONGINGKLANG
SUKHO MANOKHUN
WASAN PATAWONG

项目索引
INDEX OF PROJECTS